Jonathan Reisman, M.D., is a doctor of i[...] who has practiced medicine in the world's most remote places – in the Arctic and Antarctica, at high altitude in Nepal, in Kolkata's urban slums, and among the Oglala Sioux in South Dakota and in rural Appalachia. He heads a non-profit to improve healthcare and education in India. His writing has appeared in the *New York Times*, *Slate* and the *Washington Post*. He lives in Philadelphia with his wife and children.

Praise for THE UNSEEN BODY:

'A magnificent travelogue through the human body by a truly intrepid explorer. A genuine "must" for anyone who is even remotely curious about their own body and how it works. I really loved it'
Sue Black, author of *Written in Bone*

'A fascinating, lyrical book ... Reisman's experiences in other cultures bring a richness and depth to *The Unseen Body*. The way he thinks about the body and medicine – the rivers and tributaries, the flowing and unclogging, the top-down organisation of the brain – is extraordinary!'
Mary Roach, author of *Stiff*

'If you are fascinated by the human body and how it works, or you are thinking about studying medicine, or you are just a curious person, you will find this book a joy to read ... This is a great and easy read which I heartily (excuse the pun) recommend'
Alex Rodgers, author of *The Deep*

'The author's literary approach to this complicated subject proved excellent bedside material. [Reisman] comes across as a generous and thoughtful physician ... [who can] make sense out of it all'
Wall Street Journal

'Reisman offers a behind-the-scenes look at life itself via an odyssey through the human body ...
deep curio[...]

'Quirky, never-dull popular science'
Kirkus Reviews

'An intelligent, innovative, eclectic and accessible book in
which Reisman views the human body with a remarkable degree
of lateral thinking. Sometimes funny, often gruesome but
always entertaining and educational'
Samer Nashef, author of *The Angina Monologues*

'Terrific, memorable, original, and full of information that
informs one's understanding not only of the body but what has
stopped us from knowing more'
**Kenneth S. Brecher, cultural anthropologist and
author of** *Too Sad to Sing*

'I highly recommend this book to everyone interested in understanding
and appreciating the marvels of the human body.'
**Warren Zapol, M.D., professor of anesthesiology at Harvard
Medical School, Antarctic researcher, and inventor**

'An elegant, elegiac, and deeply enjoyable meander through human
anatomy . . . the images Reisman conjures will linger long
after you've devoured his delightful prose'
Nicola Twilley, co-author of *Until Proven Safe*
and co-host of *Gastropod* **podcast**

'A remarkable travel narrative that renders [the human body] both strange
and fascinating, so much so that the reader will feel not only delighted,
but also proud to be a lifelong inhabitant of it'
Lawrence Millman, author of *Fungipedia*

'An engaging book likely to pique the curiosity of readers interested in a
wide range of medical conditions or naturalistic medicine'
Library Journal

'Reisman engagingly relates provocative stories for the fifteen body
parts uncovered in this treatise, and goads the reader to re-evaluate their
perception of the body'
City Book Review

THE
UNSEEN
BODY

*A Doctor's Journey Through
the Hidden Wonders of the
Human Anatomy*

Jonathan Reisman

WILDFIRE

First published in 2021 by FLATIRON BOOKS

First published in Great Britain in 2021 by
WILDFIRE
an imprint of HEADLINE PUBLISHING GROUP

First published in Great Britain in paperback in 2022 by
WILDFIRE
an imprint of HEADLINE PUBLISHING GROUP

1

Cataloguing in Publication Data is available from the British Library

ISBN 9 781 4722 8941 4

Offset in 10.01/14.59 pt Adobe Jenson by Jouve (UK), Milton Keynes

Printed and bound in Great Britain by Clays Ltd, Elcograf S.p.A.

Headline's policy is to use papers that are natural, renewable and recyclable products and
made from wood grown in well-managed forests and other controlled sources. The logging
and manufacturing processes are expected to conform to the environmental regulations of
the country of origin.

HEADLINE PUBLISHING GROUP
an Hachette UK Company
Carmelite House
50 Victoria Embankment
London
EC4Y 0DZ

www.headline.co.uk
www.hachette.co.uk

To Kai and Sierra,
may they wander, explore, and be
fascinated by everything

Contents

THE UNSEEN BODY

Introduction

got my first glimpse under the human body's hood on the very first day of medical school in anatomy lab, the class in which I would dissect a cadaver. My classmates and I dug only as deep as the muscles of the cadaver's back that day, but we uncovered a captivating view into the mechanics of how we move our arms and flex our spine. That glimpse inside the body felt like a behind-the-scenes look at life itself, and before that first lesson ended, I had decided that when I died, I would donate my own body for the very same medical school dissection.

Over the following months, we dug deeper into the cadaver and examined each of the internal organs one by one—the body's hidden workhorses that labor every day of our lives to keep us healthy. Liver, stomach, intestines, lungs, heart, kidneys—each was a unique inhabitant of a new world I was discovering, and each had its own specific role in keeping us ticking. I saw that all the human body is a stage, and the internal organs are its primary players.

Taken as a whole, the human body's shape is complicated—a bulbous head, four barely cylindrical limbs, and jutting corners of bone disturbing any hint of simple geometry. But the body really has only two sides: an outside and an inside. Our outside lives begin on the skin's surface and include the everyday world of appearances and conversation, of air, nature, and other people. Most people spend their entire lives focused exclusively on the outside world, but medical training necessarily centers on the body's inside life—the one that most people ignore until some symptom draws their attention and conjures up fear of a dreaded, mysterious process happening within. Though our insides glimpse the light only in operating rooms or after terrible traumas, they are the human body's true biological business end.

For each body part, I memorized details of its structure and function, handled rubbery, preserved specimens, and studied microscopic views of its cellular architecture. I came to understand each organ's function in sickness and in health, and then regurgitated its detailed story for an exhausting multiple-choice examination before moving on to the next one. Medical school meant speed-dating internal organs, and I found myself falling in love with all of them.

I never wanted to become a doctor. Before I became an explorer of the human body, my passion was exploring the natural world. Strangely, my interest in nature was sparked while at college studying mathematics in the middle of Manhattan—I think that the oppressively urban surroundings along with the abstract unnatural perfection of math pushed me to sign up for a guided walk of wild edible plants in Central Park. On a warm summer day, we walked through the park's forests and fields, with tall buildings towering in the distance—we found a weed called poor man's pepper, which had a sharp, biting taste, and picked a handful of delicious wild raspberries. The idea of obtaining sustenance for the human body directly from nature intrigued me, and learning to identify a few species of plant opened my eyes to a whole new world.

After that first glimpse, I started learning to identify every species of plant, animal, and fungus that I might encounter in the woods, especially the edible ones. I became fascinated by the idea of living off the land, obtaining from it not only food but also materials for essential subsistence crafts. I read books about how willow branches can be woven into baskets, and how animal skins become clothing, and my dorm room soon became littered with wood shavings and half-finished projects. I longed to travel the world and visit different cultures to learn how each interacted with its own unique natural environment.

Once I graduated from college, I got my chance and developed an incurable wanderlust that began while living in Russia. By experiencing different cultures and understanding their worldviews, I broadened my own understanding of the natural world and the species inhabiting it. I spent several years adrift, sustaining my nomadic lifestyle with odd jobs at a summer camp and research grants to study native peoples of the Russian Far East. I was unsure of what I wanted to do with my life—I thought about becoming a craftsman, getting a degree in ecology, or taking up residence in the wilderness.

Ultimately, I decided that becoming a doctor was the best way to combine three of my passions—analytical problem-solving, working with my hands, and continuing to explore the world, but with the ability to offer medical help to the people I would encounter in the earth's far-flung regions. Once I started medical school and my scalpel met the cadaver's skin, I discovered that exploring the body felt quite similar to exploring the outside world—learning about internal organs reminded me of learning about species in nature. Each organ was a different creature with its own unique appearance and particular behaviors, and each could be found tucked into its own corner of the body's insides, its particular native habitat.

Once I started at Robert Wood Johnson Medical School in New Jersey, I was surprised to find that the same skills I use in observing nature had prepared me well for being a doctor. One day while exploring a small, neglected patch of woods behind the medical school's parking lot, I came upon a cluster of wild mushrooms. I had pushed through a tangle of poison ivy and other weeds ringing the lot, drawn by a curiosity about what lay beyond.

At first, I thought the mushrooms were just a piece of garbage driven by the wind into that spot. But as I walked closer, I realized that what I was looking at was alive—a creamy, light-orange collection of fungi, each one poised slightly over the moist, shaded soil. My mushroom identification skills were still rudimentary, and I wondered if they were medicinal, hallucinogenic, or one of the poisonous species I'd heard mentioned in toxicology class, a few bites of which can cause fulminant liver failure. Or perhaps they were edible. The day after I found those mushrooms, I bought a field guide so I could identify them, and I began diving into the world of edible fungi alongside my medical school curriculum.

I began filling my head with information. Most days, after exploring the human body in anatomy lab, I traded in my scalpel for a basket and delved farther into the woods behind the parking lot or roved across other promising patches of land, my sight and attention drawn to every wood-chip pile and overwatered suburban lawn in search of mushrooms. I came to covet the dazzling intuition and encyclopedic knowledge of seasoned mushroom foragers in the same way that I admired senior doctors—both showed a nonchalant certainty as they made life-and-death decisions about which mushrooms were edible and which were poisonous, or which patients needed urgent treatment and which were healthy enough to be discharged home. I wondered if I could ever make such decisions with comfort—if confidence would ever eclipse my anxiety.

As I gained experience in identifying both mushrooms and diseases, I realized that each medical case, like each mushroom, is a diagnostic puzzle. The word *diagnosis* actually means "to know apart from," or "to distinguish," and this is both the physician's and the forager's task. Whether confronting a diagnostic puzzle in a green field or a sterile hospital, my mind worked to solve it in exactly the same way—by homing in on subtle hints to tell look-alikes apart. Good observation skills were crucial for noticing the subtle differences in shape and color necessary for correctly identifying a mushroom, or for picking up a fruity whiff in the forest that tells of nearby black trumpet mushrooms, which are often smelled before they are seen. In medical school I was learning that careful observation was just as crucial for recognizing a feverish child's nostrils flaring with each breath, telling me that the diagnosis is more likely an infection deep in the lungs rather than a simple viral cold.

But in both fields, observation gets you only so far. To truly understand the body, as well as the natural world, one must comprehend ecology. In the natural world, ecology is the study of how individual species interact with one another and with the land and climate. An experienced forager knows which mushrooms to expect based on region, weather, season, and recent rainfall patterns, the sort of tree overhead and forest duff underfoot. By the same token, a physician understands that diseases have an ecological context of season and geography—doctors expect Lyme disease in particular regions during summer and influenza and carbon monoxide poisoning in the winter, and, as in foraging, knowing what to look for helps them to see it. I realized that the same combination of deep knowledge and keen observation goes into diagnosing both mushrooms and disease, and I came to revere the skills of senior physicians, who—like elders in a foraging society—are the repositories of something textbooks cannot teach.

I t was only once I left the classroom and began working in the hospital during my third year of medical school—when living patients replaced multiple-choice questions—that I came to more fully appreciate the way different elements of the human body work together to keep us alive and make us the people we are. During my surgery rotation, I got another view under the hood, but this time the body was alive. A young man who had been in a car crash needed his ruptured spleen removed urgently, and I hurriedly scrubbed my hands to prepare for the surgery. Standing beside the operating table with a sterile gown, cap, gloves, and mask to cover all but my darting eyes, I watched the surgeon slice open the patient's abdomen just as I had done to my cadaver.

But this time the wound bled. The flow of red blood told me that the patient's heart was beating, his lungs were breathing, and his blood pressure was adequate. As I helped pull open his abdomen, the living flesh felt warm in my gloved hands—it told me about the warming secretions generated by the patient's thyroid and adrenal glands. And the young man's animated bowel squirmed like earthworms, as if the surgeon and I had cut into the earth's grassy surface and pulled aside turf to expose the living soil beneath, riven with bustling creatures. His lively intestines spoke of adequate nutrition, balanced electrolytes, and proper kidney and liver function.

Opening living patients, and seeing and touching active internal organs, revealed something that a cadaver's dead, static tissues frozen in time by death and formaldehyde never could—that all our organs are inextricably interconnected in a web of mutual interdependence. A cadaver on a slab could give no hint of how organs work together in life, just as taxidermy specimens or pressed plant leaves say little about the complexities found among species in a shared habitat. In living patients, every observation I made necessarily told of distant organs cooperating harmoniously. The *organs* I had learned about one by one finally became integrated into an *organism*.

Just as each species in an ecosystem is perfectly adapted to its own ecological niche, each facet of the human body plays a specific role in maintaining the body's balance—what we call homeostasis. Ecology of the body, additionally, is the big-picture view one acquires by stepping back from an individual part to comprehend the interconnectedness of the whole. Through my medical training, I discovered that being a doctor—especially a generalist—means not only having deep knowledge about each body part but also being able to see the entire human body, as well as the person inhabiting it.

This book is a travel tale. My journey of discovering the wonders and ailments of the human body began with looking inside my cadaver, which also meant indirectly peering inside my own body, as well as the bodies of every person I would ever meet or treat. The journey continued through medical school and into my residency—the training in internal medicine and pediatrics I completed thereafter at Massachusetts General

Hospital (MGH)—and each patient I evaluated contributed to a greater understanding of these bodies we inhabit.

Once I finished residency and began working as an independent physician, my career continued developing along an unusual path. Instead of working at a prestigious academic hospital or becoming a specialist, I sought out remote and culturally unique regions of the world in which to practice medicine. From a clinic in high-altitude Nepal to an emergency room in Arctic Alaska, practicing medicine in different places showed me new cultural perspectives on the human body that deepened my own appreciation and made me a more informed physician.

This book is about parts, but also how those parts compose a whole. Each chapter offers stories about a specific body part or bodily fluid from the perspective of a physician, as well as that of an explorer and a traveler scouting out a new and unfamiliar country, experiencing its novel sights and the odd customs of its inhabitants. Medical school taught me that a human body can be broken down into individual parts, but life taught me that the body is more than just the sum of those parts.

The hidden worlds inside us deserve as much attention and fascination as the natural world. After all, the true story of the body—of the lives we live in these bodies—is the story of both.

1

THROAT

When I first studied human anatomy in medical school, one body part in particular seemed rather stupidly designed: the throat. Our throats are where swallowed food and inhaled air, after entering the body through the same mouth, diverge into their respective tubes—the esophagus and windpipe. In the throat, the esophagus's entrance for swallowed food lies immediately behind the windpipe's for inhaled air, like two straws with their open ends abutting each other and only a few scant millimeters of tissue dividing them. Diagrams in my medical textbook showed the details of the throat, or pharynx, an all-important intersection where, with each and every swallow, food and drink career directly over the windpipe's entrance, coming within a hairsbreadth of slipping in. One tiny lapse could cause choking, suffocation, and death.

The throat's perilous construction stood in stark contrast to the brilliant design of other body parts I had been learning about. The human hand and forearm, for instance, possess a wondrously dexterous architecture of bones, muscles, and tendons with the staggering ability to grasp tools or play jazz piano. And with equal elegance, the perfectly synchronized cooperation of heart and lungs delivers oxygen to all the body's distant crevices. Anatomical mechanisms always seemed intelligent, to favor life and enhance survival—but not the throat. When my anatomy professor joked about the "idiotic" arrangement of the human pharynx, I chuckled along with my fellow students.

But years later when I worked as a hospitalist—a doctor working exclusively with hospitalized patients—a significant proportion of my job consisted of battling the fallout of the throat's flawed design, and I found little to laugh about. I frequently treated old and infirm patients suffering from aspiration, the medical term for food going down the wrong pipe. Aspiration can involve choking on large chunks of food that suddenly suffocate all of the windpipe's airflow, but most of my patients were aspirating only tiny amounts of food, drink, and saliva. Instead of choking all of a sudden, this process occurred slowly but consistently over weeks and months—and often went unnoticed by patients or their caregivers. For them, the throat's failure to keep food and air separate caused breathing troubles, frequently landing them in the hospital under my care and compounding my bewilderment at the throat's tangled anatomy. But one elderly patient, Suzanne, ended up forever changing the way I thought about this body part.

Suzanne was an eighty-two-year-old woman who lived in the suburbs of Boston. She had worked her whole life as a seamstress and did not give up her career until her late seventies, when her health began to decline. She resisted the ebbing of her independence, and when her daughter, Debra, tried to move her into a nursing home, she refused. Suzanne's physical deterioration quickened, and she had several falls, but thankfully no broken bones. Debra hired visiting nurses to care for her mother in her own house, but mental decline soon came as well, stripping Suzanne of a lifetime of memories and the ability to form meaningful sentences. When caring for her at home became too difficult and unaffordable, there was no choice left but to move her to a nursing home.

I met Suzanne a month later, when she was admitted to the hospital for a mild case of pneumonia. She recovered over a few days of antibiotic therapy, and I discharged her back to the nursing home.

But a few weeks later, she was back again with a second (and more severe) bout of pneumonia. I visited her in the emergency room (ER) while
she was waiting for a bed on the medical ward, and I found her lying on a
cot, struggling weakly against padded wrist restraints that kept her from
pulling off her oxygen mask. She had the same toothy grin and gray fraz-
zled hair, but she looked thinner and more malnourished than during the
first admission—her temples were sunken and wasted, her ribs prominent
and heaving with each strained breath.

"It's Dr. Reisman again. Did you miss me?" I joked. She mumbled something incomprehensible, which I could barely hear over the hissing flow of
oxygen and the rattling of phlegm caught in her throat. She seemed less
alert than when I discharged her last—pneumonia had once again tipped
her fragile mind, already demented by Alzheimer's, into further confusion
and disorientation. I listened to her lungs with my stethoscope and heard
the sound of bubbles blown through thick porridge.

Like many other patients I have treated for aspiration, Suzanne's decline
began with coughing at meals. Before that, her mental deterioration and
advancing dementia had proceeded steadily, but when the coughing started,
her weight began to drop and her physical and mental debility accelerated.
During her previous hospitalization for pneumonia three weeks earlier, I
had discovered that she was aspirating while eating—this was the reason
for her cough.

The root of Suzanne's problem lay in the mechanics of her swallowing. Though it is casually performed hundreds of times each day by all our
throats, swallowing is no simple feat. Maneuvering food and drink safely
past the airway's entrance requires an intricate neuromuscular orchestra of
contractions and contortions of the tongue, palate, pharynx, and jaw. Food
approaches the windpipe as we swallow, and just as suffocation seems im-
minent, several different muscles coordinate to lift up the windpipe's top
end, the larynx. This timely gesture is visible from the outside as the Ad-
am's apple jerking upward, as it tucks the airway's exposed opening under
the tongue, sealing it closed against the epiglottis, a firm flap of tissue that
perfectly plugs the larynx like a manhole cover. Food and drink can then

safely sidestep the airway, traveling beyond it and into the esophagus. Once the coast is clear, the larynx lowers again and settles halfway down the neck.

Swallowing involves the cooperation of five separate cranial nerves and more than twenty different muscles. This complicated mechanism is the body's attempt to compensate for the throat's inherently dangerous anatomy, but it is a clunky and overly complex solution to a serious problem, and therefore prone to failure. This is especially the case when people talk while eating, an attempt to keep both the esophagus and the airway open at once. It is no wonder that so many people choke to death every year.

To top off all of the throat's dangers, the number one pneumonia-causing strain of bacteria in the world lives in the back of the human throat, just above the airway's opening. Poised there much of our lives and forever ready to slip into the lungs, these bacteria wait for the moment the airway's guard is let down. For healthy people, the situation poses little danger, as these infectious barbarians are handily and thoughtlessly kept at the gate. But for Suzanne, dementia had stripped her defenses, progressive debility had weakened her throat muscles, and neurologic decline had distorted her swallowing's coordination. Even her gag reflex—the throat's instinctive rejection of anything besides air headed for the windpipe—was weak and ineffectual.

She ended up in the hospital a second time because the bits of food and saliva she was aspirating had once again brought bacteria from her throat down into her lungs, where they festered and thrived, blooming into another "aspiration pneumonia." This condition is exceedingly common in Alzheimer's patients like Suzanne, as well as in those suffering from other neurodegenerative diseases like strokes and Parkinson's disease. I have seen the same infection also result when patients aspirate during a seizure or in a drug- or alcohol-induced stupor.

I examined Suzanne each day during my rounds—the early-morning task of seeing all my hospitalized patients one after the other. I placed my stethoscope against the thin freckled skin over her back and listened to her lungs, and I monitored her mental status. I knew that the next aspiration event was an ever-present possibility and could result in another pneumonia or the need for an emergent breathing tube. Or it might quickly suffocate and kill her. Because of this dangerous quirk of human anatomy, the risk of further deterioration for my patients was highest precisely when things were most tenuous.

It seemed an unfair, and rather unintelligent, design.

Over the next three days, Suzanne's breathing improved with antibiotics, and she was able to wean off supplemental oxygen. Each morning, I noted a gradual clearing of her breath sounds, and her mental status returned to her baseline—she was still confused and uncommunicative, but slightly more awake and alert. I worked with consultants from the speech-and-language-pathology service to reduce her aspiration risk: we avoided giving thin liquids like water and juice, which easily evade the guarding epiglottis, choosing instead only thicker consistencies that her clumsy throat could more easily shepherd into its proper avenue, the esophagus.

One day I chatted with Debra at her mother's bedside after finishing rounds, as a hospital health aide slowly spoon-fed Suzanne a lunch of oatmeal and thickened liquids. I explained, as I had to the families of many similar patients before, that the aspiration risk could never be completely eliminated. We could bypass her mother's throat altogether, and feed her instead with a tube permanently implanted through the skin and into the stomach. But even when delivered into the stomach through a tube, food could still reflux up the esophagus and land in her lungs, causing choking or pneumonia.

Debra insisted that her mother would not want to be dependent on tubes, as it also clearly stated in her living will, a legal document Suzanne signed when she still had sufficient mental faculties. Having watched her mother's quality of life ebb along with her mind, Suzanne's daughter was clear and certain in supporting her mother's desire to limit invasive medical interventions.

"She used to be the strongest person I knew," Debra said, her voice cracking with memories of a person who, for her, had long since disappeared. "She always told us exactly what she wanted, and what she didn't want."

As Debra and I talked, the health aide feeding Suzanne was careful to allow enough time for each swallow. Suzanne had a tendency to store food inside her cheeks, which increased her risk of aspiration. Her appetite had improved as her fever resolved, but she still coughed occasionally while eating—each cough a reminder of her continued risk.

Coughing is one of the body's crucial mechanisms for protecting against the complications of aspiration. As a propulsive method of cleaning out from the body unwanted junk that does not belong, coughing does for the lungs what sneezing does for the nose and what vomiting does for the gastrointestinal tract. A cough is a well-honed reflex already present in small infants, a preprogrammed response triggered by any foreign material touching the sensitive airways and lung passages.

Still, everybody will aspirate in their lifetime, and coughing is the body's attempt to cope with aspiration's inevitability. For healthy people, it works quite well—a coughing fit when food or drink goes down the wrong pipe handily flushes any trespassing material out of the lungs and generally eliminates the risk of developing pneumonia. Whatever remnant of aspirated material is still left in the airways is slowly brought up and out by a self-cleaning mechanism of the lungs called the "mucus elevator," another example of the body's need to compensate for the throat.

But an effective cough requires a certain strength that Suzanne simply did not have. Her frail attempts lacked the volume and reverberation of coughs robust enough to actually move mucus. Coughing relies on chest and abdominal muscles squeezing tightly while vocal cords in the larynx are sealed closed to block the outflow of air. This builds up pressure inside the chest. When the vocal cords suddenly pop open, the pressure releases and pushes out any foreign goo. But Suzanne could not muster the strength. All of her body's defenses were overwhelmed and failing.

On Suzanne's fourth day in the hospital, my pager squawked toward the end of rounds, a nurse asking me to come quickly. I arrived at her room to find Suzanne struggling to breathe, the muscles of her neck and chest straining to huff oxygen through the face mask the nurse had placed back over her face. She had aspirated during breakfast—an event I had been dreading but perpetually expecting. Her daughter stood at the bedside teary-eyed, stroking her mother's thin veiny hand and wiry hair.

Beside Suzanne beeped a bedside monitor, and its flashing display showed a concerningly low oxygen level. Leaning Suzanne forward with the help of her nurse, I listened to her lungs—the gradual improvement in clarity over the previous days was erased, and her breathing once again sounded like bubbling gruel. With a plastic catheter, I suctioned remnants of breakfast from her throat to prevent further aspiration. The next step for a patient like Suzanne, with a low oxygen level despite breathing pure oxygen through a mask, would be intubation, a plastic tube snaking through her throat and into the trachea so that a ventilator could breathe for her.

Turning to Suzanne's daughter, I hastily rehashed our previous discussion about breathing tubes: we could call the hospital's intubation team and transfer her to the intensive care unit (ICU), or we could focus on her comfort and avoid such heroic measures.

"Mom would not want to be kept alive on a machine," she stated firmly.

I ordered the nurse to give a medication to relieve the panic of suffocation, a measure of comfort that was within her well-defined goals of care. The nurse left the room and returned with the medication, a few milliliters of clear liquid inside a plastic syringe, and injected it through the IV in

Suzanne's arm. Within minutes, the look of fear in her eyes calmed—her agitated breathing pattern slowed slightly but persisted.

As a physician, I was trained to fight unceasingly against disease and death, to battle perpetually against the human body's failings— including aspiration. But for certain patients, the time comes to shift priorities. Such decisions are difficult and always made privately between patients and their families, in consultation with their healthcare providers.

Bodily decline comes in different flavors, and the choice to accept or forgo medical interventions is different for every patient. For Suzanne, Alzheimer's had left her with no mind to watch over her failing body, and she had decided years earlier that she would not want to live long with such a life. Other patients value extending life's duration at all costs, especially when their mental faculties remain intact. For example, a close family friend was suffering from amyotrophic lateral sclerosis (ALS), a devastating neurological condition in which a person's muscle strength gradually and irreversibly degrades into total paralysis, but with no accompanying mental decline whatsoever. Unlike in Alzheimer's and other forms of dementia, my family friend's mind remained perfectly sharp as the body around it wasted away. When he began to aspirate constantly and could no longer eat through his mouth, he chose differently than Suzanne and had a feeding tube placed into his stomach.

Soon breathing became difficult because he couldn't even muster the bare physical exertion of drawing air into his lungs—a sign of terminal weakness. He opted for a breathing tube permanently placed through a hole in the front of his neck. For this friend, such interventions would prolong his life but not slow his physical decline. But they offered value to him as long as the mind trapped in his failing body could still enjoy the presence of his grandchildren. In the last days of his life, he sat completely paralyzed in a wheelchair, but still could relish the feeling of being hugged and kissed by his adoring family.

When I worked as a hospitalist, conversations about a patient's goals of medical care were an everyday part of my job. In many of my elderly and aspirating patients who could no longer communicate their wishes for or against aggressive therapies, legal documents often conflicted with a relative's bedside opinion, or family members disagreed among themselves about the options. More than once I've seen an elderly patient's adult child who had been estranged for years return to their parent's deathbed with unrealistic hopes of recovery. Another adult child who had been at the parent's side through years of decline, who had accompanied them to innumerable doctor's office visits and through repetitive hospitalizations, often was the one who knew it was time to let go.

A dying patient frequently exposes unpleasant dynamics within families. I have often been in the middle of strained sibling relations degraded over decades that finally came to a head like an abscess under pressure bursting forth with thick, foul-smelling pus. Just as the human body and mind degrade with age, human relationships also begin to creak and fray over years. Personalities harden like arteries, and the bustle of life strips people of the energy to work through the inevitable accumulation of misunderstandings and slights. As the hospitalist—and a complete stranger—my job was usually to mediate and soothe, to ease familial tension just as I eased the pain and suffering of a human body as it exited life.

Besides the intake of food and air, the throat is also where exhaled air from our lungs is whipped into voice by the larynx. It is through our throats that we express ourselves and how strong-willed and independent people like Suzanne make their wishes known. As a physician, my job is to listen to patients, especially when they can no longer speak for themselves. People who fear being victimized in their last days by healthcare workers and the painful bodily invasions of medical care should sign living wills dictating which measures are off-limits. Research shows that when doctors themselves become patients at the end of life, they more readily wish to forgo aggressive therapies, probably because we often witness, and personally enact, the senseless brutality that comes at the expense of quality and comfort. Throughout life, each person speaks for themselves, but when our bodies and voices fail, we must rely on our families and friends, or documents like a living will, to speak for us.

Suzanne's daughter was atypical in her keen understanding of her mother's inevitable trajectory—and the fact that the living will reflected her feelings made my job easier. Caring for Suzanne was one of the most tension-free end-of-life experiences I had yet had with a patient, and it left an indelible mark on me as a young physician. Debra had been deeply involved in her mother's care over the years, and she knew as surely as I did that death was coming soon no matter what we did.

Suzanne survived the initial aspiration event, but over the following two days, her temperature rose into fever once again and her oxygen level declined further, a sign that her throat's bacteria had again breached her lungs and were burgeoning into another pneumonia. Her condition worsened despite powerful antibiotics.

Later that day, Suzanne pulled the oxygen mask off her face, as she had on admission, but this time we let her. In her throat, mucus accumulated and began to rattle with each breath, a sign that the last strands of her body's defenses were unraveling. As Debra and Suzanne's large extended family surrounded her hospital bed, her breathing gradually sped up, and then progressively slowed to a final halt.

During embryology, the human body's design is forged microscopically in the womb. We each begin fetal life as a flat, microscopic disk of cells that, just a few weeks after conception, rolls itself, crepelike, into a tube. This roll creates the human body's basic blueprint: a tube with an entrance at one end and an exit at the other. The body grows and embellishes itself from that starting point, its structural complexity soaring. But the original tubular configuration remains throughout life. As grown humans, we are little more than extravagantly decorated tubes with entrances for food, drink, and air at the front, and all our exits clustered at the other end.

This tubular construction is the origin of the throat's design. As each human embryo develops, the body's single common entrance at the front end splits into two respective tubes side by side—one for food and one for air—and the enduring peril of choking and aspiration is born. The body compensates by sprouting a face and brain to better regulate what is allowed through the entrance, and develops protective mechanisms like swallowing, coughing, and gagging, which work most of the time.

Beginning with the first breath after birth, air and food are precisely partitioned in our throats—a lifelong pharyngeal juggling act. Keeping food out of the airway is among our body's most basic responsibilities, but when the throat of a person debilitated by strokes or Alzheimer's can no longer sustain the coordination of juggling, aspiration is how the body gives out. It is among the most common causes of death in such patients.

As a medical student, I mocked the throat's design, and as a young hospitalist, I viewed it as a significant threat to my patients. But after years of working as a physician, I came to also see it as the body's way out of life. Aspiration pneumonia was once called "old man's friend" because it often brings a dignified end to prolonged suffering in the elderly and ill. Like a cyanide pill kept in a locket, the throat's precarious anatomy becomes a degraded body's escape hatch. And sometimes, no matter how much we want the people we love to hold on for a little longer—such as when my wife's grandfather with Parkinson's disease was repeatedly admitted to the hospital with one pneumonia after another—the end cannot be forestalled any further. Sometimes the body decides for us.

Through caring for patients like Suzanne, I realized that there is wisdom in the throat's design after all.

2

HEART

The human body is a deep and labyrinthine cave riddled with channels and conduits, and through each flows some type of bodily fluid. The body's expansive catalog of ailments that I studied and memorized basically amounts to all the ways in which the flow of those internal fluids can be disrupted. Stoppages of urine flow cause kidney failure and urinary tract infection; blocked middle ear drainage leads to ear infections, and the same goes for congested sinuses; inadequate outflow of mucus from the lungs leads to pneumonia, while stony concretions formed in gallbladders, kidneys, salivary glands, and the balance centers of the inner ear obstruct the flow of their respective fluids, causing untold woe; appendicitis, diverticulitis, abscess, and constipation—all result directly from a clog stopping up the flow of some fluid sloshing through the body's corporeal pipes.

The necessity of flow resembles a tenet of traditional Chinese medicine, where a stoppage in the flow of chi is seen as a cause of most illness. But I learned that Western medicine has the same overarching principle: human health depends on the steady, consistent movement of bodily fluids, and a physician's task in treating disease is to alleviate blockages and allow fluids to resume their proper motion. In other words, most of the practice of medicine is plumbing.

One of the body's deadliest plumbing problems is a heart attack, and the first time I ever made this diagnosis by myself, I was in a rush to get home. It was just a few months after I'd finished residency, and I was working as a newly independent physician in an urgent care center outside of Boston. It was after a long and busy day, and I was tired, so I grumbled to the nurse when a middle-aged couple walked in the door just before closing time.

Jared would be my last patient of the day. Hoping to get through the visit as quickly as possible, I followed him into his bay and introduced myself, asking what had brought him into the clinic. Early that morning, he told me, he'd developed an odd feeling in his chest. He initially shrugged it off as acid reflux and resisted his wife's pleading to get it checked out. The odd feeling persisted, then eventually worsened into actual pain that began moving from his chest toward both his shoulders. It wasn't until later that afternoon that he finally acquiesced to his wife's demand and sought help at my clinic just down the street from his house.

As the nurse placed electrocardiogram (EKG) stickers with attached electrodes in the usual pattern across the left side of his chest and on both his arms and legs, I asked more questions about his pain to better determine its cause: Did it feel like a sharp knife, like a burning sensation, or like pressure? Did it get worse with physical activity? Was there any associated sweating or shortness of breath?

Chest pain is among the most common reasons people seek medical attention, mainly because it is the single symptom most associated with suddenly dropping dead—but most instances of chest pain are not caused by serious or fatal conditions. I hunted for clues that would help me differentiate the mortal threat of a heart attack from more benign causes of Jared's symptoms, such as acid reflux or a muscle strain in the chest wall. Some of his responses worried me that it could be a heart attack: he said the pain worsened when he walked up stairs and improved with rest, and the pain moving from his chest to his shoulders was particularly concerning.

As he spoke, his wife glared at him, a look I've since seen on the spouses of many stubborn patients.

Within three minutes of their arrival, the EKG machine churned out a recording of the voltages that coursed through Jared's heart and sparked its muscle into contraction with every beat—a tracing of black squiggles across a strip of shiny red paper. I grabbed it and held it close to my face, studying the oscillations of electricity.

I froze, and my eyes widened. The EKG showed subtle but clear signs of a heart attack.

Not wanting to share the alarm suddenly written all over my face with Jared and his wife, I kept my gaze down at the paper. I traced the EKG with my eyes again, doubting myself. During residency, I had become accustomed to having a more senior physician oversee my work, confirming or refuting my initial impression of a patient's diagnosis. Now that I was on my own, my brain hesitated to come to the definitive conclusion staring me in the face. I knew that making the diagnosis of a heart attack would immediately lead to a cascade of events: I'd have the nurse give him an aspirin and place an IV in his arm; I'd notify a cardiologist at a large hospital in Boston who would prepare a cardiac catheterization unit requiring the mobilization of a team of five or more people; I'd call for an ambulance to rush him to the hospital; and, as an unavoidable and dreaded first step in this cascade, I'd scare the shit out of Jared and his wife.

The whole sequence ran through my mind before I finally looked up. "You're having a heart attack," I said.

The anger immediately drained from his wife's eyes and fear took its place.

I was taught to be blunt and clear when giving a diagnosis—not to waste time or hide behind abstruse medical terminology—but inside I felt unsure and nervous. Even though the EKG's message was obvious, some small part of me still feared that I was completely wrong. As a doctor, my job is reporting news about the state of internal organs hidden inside my patients, and I knew that displaying confidence, regardless of whether I actually felt it, was crucial to delivering the news with certainty to the people who most deserved to know.

As we waited for an ambulance to pick up Jared and transport him to the hospital, the nurse placed him on the cardiac monitor, a continuous form of EKG that displayed his heart's electrical pattern on a screen. Heart attacks are among the most serious diagnoses, but I watched the screen carefully for an even more dire emergency that could arise at any moment: cardiac arrest. Both are medical emergencies, but they represent completely different problems of the heart.

A heart attack like Jared's is an issue with plumbing. It is caused by a blood clot no bigger than a stub of pencil lead forming in one branch of the coronary arteries, blocking blood from delivering oxygen and nutrients to a portion of the heart's muscle.

Cardiac arrest, on the other hand, is an electrical problem. When it occurs, a patient's heart arrests and ceases its beat, and determining the diagnosis is as simple as detecting no pulse in the patient. Confusingly, heart attacks can lead to cardiac arrest—when heart muscle is starved of blood flow, it becomes irritated, and its normally coordinated electrical system can be disturbed. Sometimes there is enough localized electrical chaos for the heart's entire electrical rhythm to spiral out of control, and the heartbeat evaporates all because of one faulty area of the organ.

The heartbeat is the rhythmic foundation of the human body's life, so patients in cardiac arrest are technically dead. They still have a chance of being revived, through cardiopulmonary resuscitation (CPR) and electric shocks. The chest compressions of CPR are a method of pushing blood from the heart and beating in the stalled heart's stead, and electric shocks can restore a normal electrical rhythm like jump-starting a car. And many patients survive. But once the chances of restarting someone's heart seem more and more dismal, I make the call and shout, "Stop compressions!" I then read aloud the time displayed on a clock on the wall, because the moment we give up on the heart is the moment that death is officially declared.

In heart attacks like Jared's, despite the heart's partial suffocation, it still continues to beat normally (which is why he was awake and able to converse with me, answering my myriad questions about his pain). In heart attacks, minutes matter, but in cardiac arrest, it's the seconds that count. Cardiac arrest, also called a "code blue," is the most urgent emergency in all of medicine, which is why it is the only diagnosis announced over a hospital's overhead speakers. When other organs stop working, death usually follows in minutes, hours, or days—the body can sometimes live for years after brain death—but when cardiac arrest occurs, death technically happens in that same instant. The death of our hearts *is* death.

Jared's monitor would sound an alarm if his heart's urgent plumbing problem devolved into an even more emergent electrical problem. If it did, the nurses and I were ready to start pounding on his chest and charging up the defibrillator for a zap.

The fact that a tiny problem with the heart can kill us in an instant further shows this organ's preeminence in our bodies. For centuries before physicians understood how heart disease kills people, the same primacy of the heart was the stuff of poets, lovers, and soul seekers. Even today, despite dramatic advances in knowledge of the body's anatomy and physiology, we still draw cartoonish, bilobed symbols of the heart in love letters and text messages—an anatomically inaccurate design that has more to do with Valentine's Day advertising than human biology. Such symbols may be cute, but the actual human heart more strongly resembles an oversized avocado.

We speak as though love and passion spring from this organ as surely as blood surges from the left ventricle into the aorta, but here we might be giving too much credit to a simple mechanical pump. Linking emotion with the heart comes from a time before we truly understood it, before humans knew of electricity or learned that the heart runs on it. I understand why people make this connection—when I was smitten with one of my fellow students in medical school, it felt like a burning heaviness in my chest that certainly could have been actual heartache. To the ancients, the bloodred heart probably also seemed a likely source for an emotion that might flush our cheeks with the same color.

Though we now know better, we still hold on to older ways of understanding our organs that have lived on as poetic interpretation. Ram Dass, the spiritual teacher, implored his students to live in their hearts rather than in their brains. While the busy brain offers only a wordy monologue full of constant judgment, Dass teachings focused on the heart, which he saw as the seat of a far deeper awareness. He called this wordless core of our bodies the "heart center."

Anatomically and physiologically, this makes sense: the heart sits in the very middle of our chests, sandwiched between two framing lungs that continually inflate and deflate. There it beats for as long as we are alive, a fist-sized hunk of hollowed-out muscle, with blood surging through valves that open and close to control the flow. The heart is more central to the maintenance of minute-to-minute bodily health than any other organ.

It is also the only internal organ whose primary function is self-serving. The heart's rhythmic contractions pump oxygen-rich blood through arteries in order to nourish every cell in all of the body's distant crannies. But it also pumps blood to itself through the coronary arteries in one of the body's few self-referential loops. Like a snake biting its own tail, the heart is the ouroboros of our internal organs, and a useful symbol of introspection and self-reflection.

Another instance of such a loop occurs when the brain thinks about its own function. And however much we exalt our brains, the heart is still a more primal organ, since its beat is what permits the brain to function at all.

Geographically and functionally, the heart *is* our center.

Jared left the urgent care clinic in an ambulance that transported him to MGH, where a cardiologist awaited his arrival. If the flow of blood through his blocked coronary artery was not quickly restored, the cardiac muscle cells in that area would likely die and permanently weaken the heart's ability to pump blood. Each medical specialist knows intimately the plumbing of their organ, and cardiologists spend their careers studying the heart's own fluid mechanics. Those subspecializing in catheterization are trained in performing the actual unclogging. By squirting dye into the coronary arteries and taking a series of X-rays, they reveal a ghostly apparition of spidering blood vessels crawling over the heart's surface and can directly visualize the blockage. These cardiologists then insert a long, flexible wire called a catheter through an artery in the groin that tracks several feet up to the heart.

I later learned that the X-ray images during Jared's catheterization showed the dye stopping dead right in the middle of one of his coronary arteries. The offending clog was found. To open the sluice, the cardiologist threaded a wire directly through the blockage and sucked out the blood clot, reestablishing blood flow in the same way a plumber snakes a drain.

Within an hour of Jared's describing his chest pain to me, he had two stents in his coronary artery to unblock the flow of blood and to keep it open. While I ate dinner at home, blood flow had finally returned to the portion of his heart that had been suffocating, a kitchen faucet sputtering and gurgling back to life after its water supply is turned back on.

To maintain the flow for the long term, Jared's doctors prescribed several new medications. A daily aspirin along with Plavix (clopidogrel), another blood-thinning medication, would prevent blood from clogging on its way to his heart, and a statin would prevent the further buildup of gunk within plumbing pipes. Along with diet and exercise, these medications—the equivalent of Drano for his coronary arteries—might help clean the buildup from his artery walls and fortify them against future blockages.

Before discharge, Jared was also counseled on depression and anxiety, which are exceedingly common in patients after suffering a heart attack. Such an event can be a watershed moment in a person's life. Survivors often face new physical limitations and live in fear of a second heart attack, both of which contribute to psychological distress. Just as the human body's health depends on continuous flow, the same is true of the mind—when fear, sadness, and worry are bottled up and kept inside, a crystal clear stream turns into a stagnant swamp of isolation and anger. And without good plumbing, mental health deteriorates—and interpersonal relationships along with it. Jared was scheduled for talk therapy with a counselor the following week; the cure for psychological problems often requires simply talking, opening the sluice of communication with fellow human beings, and restoring a healthy flow.

Before becoming a doctor, I traveled the world, and my favorite part of flying in airplanes was the close-up views of land shortly after takeoff and before landing. From the bird's-eye view of a passenger jet, I enjoyed studying the earth's surface—the clustered buildings of towns interspersed with orderly farm fields and patches of forest, and an occasional factory smokestack belching smoke upward at me. But more than anything, I admired the waterways breaking serpentine paths across the earth's surface, writhing and flagrantly disrupting the humanized geometry of farm and town.

From far above, I could see the branched and coalescing structure of river systems. The tiniest rivulets of water converged into larger streams, and those streams again united with others while steadily growing in size. No stream flowed in isolation for long, each soon becoming a tributary itself. The fractal pattern of branching flows was universal, from high in the mountains to the seashore. When my airplane flight coincided with sunrise or sunset, the sun's slanted rays accentuated the terrain below me, and I could see more clearly the correlation between the shape of land and how water flowed across it. Each section of land with its system of rivers was a watershed, a geographic area draining into a single common waterway.

Even before I went to medical school, I recognized the same pattern on my own body. My forearm, perched atop an airplane armrest as I craned my neck to see out the small porthole window, was emblazoned with blue veins squirming across its surface much like the watery squiggles on the living map far below. Creeping up my arm, they combined into larger blood vessels like streams draining a terrestrial watershed.

On one trip to the Russian Far East, I explored waterways more intimately—by traveling along them. While in Kamchatka conducting anthropological research on indigenous peoples, I joined a local family on a weeklong horseback trip to their summer hunting cabin. Vasily, a short thirty-six-year-old Even man with prominent cheekbones and stubble on his chin, led the way on his small horse. His wife, Olga, followed behind, and their five-year-old son, Andre, his cheeks chubby and ruddy, sat behind his mother on her horse and held on to the folds of her jacket. I rode last on my own horse as we started up a large valley out of Esso, the largest town in central Kamchatka. We traveled along the Bistraya River, moving upward against its white churning flow as we left town. Andre often turned around on his horse to make silly faces at me as we rode.

A few hours after setting out, we came to a confluence where a smaller tributary branched off from the Bistraya. We turned right and followed the smaller river for a while, and we camped along it that first night, pitching our tents on its bank and using the stream to boil water for tea and dinner. After packing up our campsite and moving farther upstream the next day, we came to yet another confluence, where an even smaller river branched off the one we had followed. Again, we turned and rode along the smaller tributary, climbing ever upward into the mountains as we traced smaller and smaller streams. Vasily knew the route by heart and told me the name of each waterway as we went. Eventually our guiding stream lessened into a mere trickle, one that, Vasily said, had no name. It then disappeared for good beneath rock-strewn tundra as we crested the valley's headwall and stood directly on a mountain pass.

From the pass, I could see over the other side into another immense valley dropping away abruptly with its own trickle of water flowing downhill in precisely the opposite direction, a mirror image of the valley we had just ascended. I could see how the pass formed a boundary between two watersheds, and that the next stream far below represented the tender, trickling beginnings of another system of branching and coalescing streams. We followed that next waterway down from the pass, and over the subsequent days it continued to join with other tributaries and to grow in size, and I began to viscerally understand the shape of watersheds on the land. I also saw how Vasily's intimate knowledge of the watershed's contours was essential for navigating through such remote and roadless mountains.

A few weeks later, while sitting along the salmon-choked Chailino River in northern Kamchatka, I watched the water flowing and the fish fighting their way upstream, and my own thoughts about my future gathered and coalesced. My trip to one of the earth's most remote and beautiful places was coming to an end, and I imagined one day visiting again, not as a tourist but as a physician able to help the region's people, who had been so gracious and hospitable during my stay. It was on that river that I made the decision to apply to medical school when I returned home, a watershed moment of my own.

After I was accepted to medical school and began learning human anatomy, I discovered watersheds everywhere in the body. Bile drains from the liver into tiny ducts that join into larger and larger flows, eventually joining with the pancreas's own branched flow of digestive juices before delivering the combined payload into the small intestines. Similar patterns of flow exist in the drainage of salivary and mammary glands, and I memorized all of the human body's terrain maps and branching conduits for bodily fluids.

The map of the cardiovascular system was the one most forcefully drilled into my memory, and it reminded me of what I had learned during my travels in Kamchatka. The path we traveled up the Bistraya River mimicked the travels of every blood cell that leaves the heart on its way to all of the body's tissues. Each drop of blood pumped out of the heart first enters the aorta, the body's main river. When it reaches a confluence, it turns and follows the smaller tributary artery, which leads toward smaller and smaller branches of the arterial tree, each turn bringing blood closer and closer to its destination in the body.

While studying the body's smallest blood vessels, the capillaries, I recalled standing with Vasily next to the tiniest nameless stream near the mountain pass. While every large and medium-sized blood vessel in the body has a name, the smallest capillaries flowing deep in the body's remote hinterlands remain unnamed, even in the most detailed medical textbooks.

The capillaries that deliver nutrients directly to the doorstep of our cells are the mountain passes of the body. It is through them that blood travels from the watershed of arteries into the mirroring watershed of veins for the return trip. The second leg of their circulatory travels begins in the tiniest venous streams that fuse and grow into larger and larger vessels. The blue-tinted veins snaking up the forearm are formed by the joining together of the tiniest rivulets trickling out of the hand and its flesh watershed of muscle, bone, tendon, and skin. Every inch of the body is necessarily part of some watershed of blood drainage, just as every drop of rain falling on land eventually finds its outlet into a stream. And each drop of oxygen-depleted blood flowing out of the human body's most distant geographies drains eventually into the body's largest venous rivers, or venae cavae.

I learned that blood vessels in our bodies are dynamic like waterways on the earth, and can modify themselves to grow around chronic blockages. People with coronary artery disease often compensate over the years by sprouting new circumventing paths through which blood can flow to cells. When a rock slide newly dams up a stream, the water eventually finds a new route around it, just as the body's cardiovascular system carves new channels for blood flow and leaves old ones to senesce as oxbow lakes.

As I studied the names, trajectories, and branch points of almost every blood vessel coursing through the body, I strived for the level of familiarity with which Vasily knew the journeying waterways of his home mountains.

I n the grand scheme of the human body's plumbing, internists—practitioners of internal medicine—focus on clogs while trauma surgeons deal mostly with leaks. Pressure is a necessary ingredient in any plumbing system, including the body's, and a leak—namely, bleeding to death—causes a drop in pressure throughout the system.

The human body is like an apartment building, with each of its trillions of cells as individual apartments needing a steady supply of fresh potable water delivered under pressure, as well as a drainage system for used waste-water that flows by gravity alone. And the pressure must be high enough to spray strongly through every faucet and showerhead in each apartment. Cells of the body similarly need a steady supply of fresh oxygen-rich blood, and arterial pressure must be high enough to distribute blood everywhere, including to the brain, the body's penthouse apartment. This is precisely the heart's service to the entire body—its contractions forcefully push blood into the arteries and thereby create the system's pressure. But when blood seeps out of cardiovascular pipes, its task becomes impossible.

Any spot on the body can spring a deadly leak, whether a bleeding stomach ulcer or an amputated limb, but the most difficult of all to patch is one in the heart itself. I once watched a trauma surgeon sewing a bullet hole in a patient's heart. The young man was covered in blood oozing from holes in his chest and back, with no pulse or discernible cardiac activity. His skin looked ghostly gray—with the pressure behind all of his body's faucets dropping, blood flow through each, including those that provide skin with a rosy color, had slowed to an ineffective dribble. There was no time to get him to an operating room, so the trauma surgeon worked directly at the bedside.

Suspecting a direct hit to the heart, the surgeon quickly cut open the patient's chest with a long arcing slice between two ribs on his left side. He spread the ribs apart and opened a view into the man's chest, and I could see the heart itself profusely leaking blood through a bullet hole in its red meaty wall. It was a rare event to directly see a heart in the ER—surgeons prefer to expose it to air only in the controlled environment of an operating room. But with no pulse, seconds mattered—just as in cardiac arrest. Cutting open the dying man's chest was the last resort in order to reach the inaccessible heart itself and plug the leak.

As his assistant steadied the injured organ, the surgeon stitched closed the bleeding hole. But when he finally got it sealed, he looked at the heart's underside and found three more holes, each larger than the initial one. A blood transfusion was being emergently infused through the patient's IV, but it was leaking out through these holes, his body a sieve putting up little resistance. A leak directly from the heart means that no matter the force with which it beats, it can no longer generate enough pressure to feed every part of the body—the patient's body was in a state of cardiovascular collapse.

Believing the patient's heart to be beyond salvage, the surgeon gave up, and the patient was declared dead at that moment. As hunters know, a direct hit to the heart is the most effective kill shot—a tiny bullet hitting the body's most central organ is the quickest way to ensure the death of even the largest creatures. In the same way, a minuscule blood clot in the coronary artery can bring down an entire human. And not all problems with the body's plumbing system can always be repaired.

I decided to pick a plumber's brain to see what additional insights I might gain about the human body. Richard Blakeslee, the head of maintenance in one of the hospitals where I have worked, fulfills the roles of both cardiologist and trauma surgeon for the hospital's plumbing system. And he solves problems the same way doctors do.

When he received a call one day that the flow of water through faucets had stopped completely in one of the hospital's wings, he immediately radioed to his colleagues elsewhere in the hospital to ask if the same thing was occurring in other wings. He explained to me his rationale: if water flow to the entire hospital had dropped off simultaneously, the solution would have been easy—he would call the borough, since the problem is in the town's delivery of water to the hospital itself.

But Blakeslee's colleagues radioed back that water flow remained strong in every other part of the hospital. Instead of being a hospital-wide, global phenomenon, the problem was regional, affecting only one branch of the hospital's plumbing system. He knew by heart each water supply pipe in the hospital, its trajectory and branch points, as well as the watershed of faucets that each feeds. The familiar geographic map of branching, interconnected flows told him where a blockage would have to be.

He headed for the hospital's valve room and walked directly to the location where the map in his head told him the problem would lie. At the exact spot where the pipe supplying the affected hospital wing branched off the main supply line, he found the cause—a valve had malfunctioned and the gate controlling its flow had dropped, sealing it off completely like the sudden blood clot of a heart attack. Replacing the valve solved the problem like a cardiac stent restoring flow to the affected hospital wing.

One day after my shift, Richard took me through a door from the stairwell I had never noticed before—we had entered the hospital's valve room. He shouted over the loud and rhythmic whirring of motors as he pointed to filthy white pipes snaking all over from floor to ceiling, enlaced with valves bearing large red wheels for opening and closing them. It was the hospital's heart, with pipes fanning out to bring fresh water to every corner of the hospital, and it reminded me of the crown of large, crisscrossing blood vessels that arcs from the heart's top side.

Watersheds of the body, like watersheds in hospital plumbing, comprise an organic and connected system, and I use precisely the same understanding as Richard when I deal with clogs in the body. When a clot forms within one branch of the coronary arteries, it blocks blood flow to only those cells fed by that one branch; the rest of the heart continues to receive a healthy flow. But this is not so easy to figure out. The heart is hidden behind an opaque chest wall made from the wattle and daub of flesh and bones, and I cannot simply radio to colleagues to test faucets in its various regions, or walk into the valve room for a look. This is where EKGs come in handy— they are a way of peering into the heart and testing faucets within it. And interpreting EKGs requires an understanding of watersheds.

The twelve electrodes on a standard EKG record the electric impulses coursing through the heart's muscle, and each represents a different wing of the organ. When blood flow to cardiac tissue is halted, the resulting disturbance in electrical activity shows in the EKG's squiggles, but the important thing is to determine whether the problem is global or regional. Every time I look at the EKG of a patient with chest pain, my first question is the same as Richard's: Is there a geographic pattern to the disturbance?

An EKG's message of heart health is not always so easy and obvious as in Jared's case. To interpret its more subtle clues, I use the coronary map that I memorized in medical school. Each region of the heart is served blood by a particular coronary branch, and when I see electrical disturbances on an EKG clustered in electrodes all representing the same wing of the heart, I become concerned. When the problem is regional, contained within a single watershed of coronary blood flow, it means that a patient's chest pain is likely being caused by a heart attack, rather than some other less serious cause.

As in Richard's problem-solving technique, and likewise for Vasily and Olga traveling through mountains, intimate familiarity with the terrain and its branching water flows is necessary. Cardiologists who perform catheterizations accomplish a similar feat when they thread a catheter through coronary arteries toward a heart attack's culprit clot—at each branch point, they know which way to turn as they climb into smaller and smaller blood vessels, eventually reaching the spot where dye stopped dead. To solve the most vexing plumbing problems, treat the deadliest medical conditions, and navigate the most remote and wild mountains, plumbers, physicians, and mountain travelers alike must take a step back from the edge of a single stream and look at the bigger picture of interconnected flows and branching drainage channels. They all must understand watersheds.

Since meeting Jared, I have diagnosed many heart attacks, and I have become comfortable setting the cascade in motion without hesitation. After years of practice, interpreting EKGs feels like navigating a familiar geography I have been through innumerable times before.

Ultimately blood flow in our bodies is different from both the earth's waterways and a building's wastewater pipes in one key respect. Water draining off the land combines into the largest rivers that reach the end of their journey at the sea in braided deltas. Likewise, effluent from a building's plumbing system flows into the largest pipes, eventually reaching their terminus in a wastewater treatment plant. But blood is different—beginning in the heart and flowing through blood vessels that branch to infinity to meet all our cells before coalescing back together again, blood eventually reaches its delta in the very same singular place where its journey began—the body's true center, the heart.

3

FECES

O f all the physical exams I learned to perform as a medical student, the rectal exam was the most daunting. I had seen it done before, and while it seemed technically easy for the doctor and only moderately uncomfortable for the patient, I dreaded the thought of broaching the subject with a patient as much as I dreaded the act itself. My fellow students and I joked often about this rite of passage, which none of us were looking forward to.

I knew eventually I would have to perform the exam, but my rite came sooner than I expected—it was while working in the hospital ward during my internal medicine rotation as a third-year medical student. An elderly woman had come to the ER the previous night because of nagging chest pain, and when the initial testing proved inconclusive as to whether she was having a heart attack, she was admitted to our inpatient service for further monitoring and tests. While reviewing her labs in the morning, I discovered that her hemoglobin level had dropped from ten to eight, indicating an extreme and sudden anemia that needed immediate attention. I reported my finding to the resident overseeing my work that month, and she sighed— another task added to an already busy morning.

"It could be a GI bleed," she said, referring to the possibility that blood loss from the patient's gastrointestinal (GI) tract could be responsible for the hemoglobin drop.

"She needs a rectal exam. Let me know what you find," she said curtly as she turned back to typing her notes on the computer. My time had come.

I knocked gently on the patient's hospital room door before letting myself in. She lay in her bed with tired eyes and gray hair still messy from sleep. This poor woman, I thought to myself—not only was she possibly having a heart attack and a GI bleed, but her first interaction this morning was going to be my first rectal exam. I was relieved to see that the room's other bed was unoccupied—nobody else would be around to witness my bumbling efforts and our shared humiliation.

"Your hemoglobin is lower today than it was yesterday," I said. "We're worried you might be bleeding from your intestines, so I need to check . . ." I trailed off. All the next words that came to mind—*anus, rectum, stool, poop*—felt incredibly silly, and I hesitated to utter any of them in conversation with this nice older woman whom I had never met before. I looked down at my hands, which were holding the packet of necessary surgical lube, and finally blurted out: "I need to do a rectal exam."

I was sure she would refuse or, at least, argue vehemently.

"Whatever you need to do," she replied, sounding slightly miffed. Before I could instruct her to do so, she rolled onto her side, making me wonder if this was not her first time.

"Now drop your drawers." As the phrase left my mouth, I cringed at how idiotic it sounded. She struggled to pull down her underwear, so I helped. "Bring your knees up toward your chest," I instructed, hoping to optimize my angle. Kneeling on the floor behind her, I put on a pair of gloves and pulled out a small paper card from the pocket of my white coat. My task was to reach my finger into her anus, pull out a smear of stool, and dab it onto the card to test it for hints of blood.

I knew the anatomy of the anus well—its puckered outlet, the multiple layers of sphincter beyond that, and the ruffled lining at its inner end, which my finger would traverse in order to reach the rectum, where stool is stored. But when I hoisted her flabby buttock, instead of the anus I was expecting,

I saw a cluster of miniature flesh-colored mountains where her anus should have been—it was the worst case of hemorrhoids I have, still to this day, ever seen. With a blob of gooey lube on the tip of my finger, I navigated through the bouquet of hemorrhoids and found the entrance where my finger easily slid in. I could feel stool with the consistency of dense clay in her surprisingly cavernous rectum, and when I withdrew my finger, its gloved covering was coated with tarry feces as black as coal.

A rush of excitement swept over me: I knew right away it was melena—black stool that signals bleeding from the upper gastrointestinal tract, usually from the stomach. I had answered many multiple-choice questions about melena over the last year, but this was the first time I'd seen it with my own eyes. It was stickier than I had imagined. I smeared the stool onto the card, and when I dribbled a few drops of clear liquid from a tiny bottle I kept in my white coat pocket, the smear of feces turned blue, confirmation that the resident had been right all along: the patient was bleeding from her GI tract.

While delicately removing both my gloves in an effort not to contaminate anything, I hurriedly explained to the patient my findings and concerns. I washed my hands with soap and water and nervously thanked her—for what, I am not sure—before rushing out of the room.

When I reported my findings to the resident, she ordered a strong intravenous antacid for the patient, a recheck of her hemoglobin every six hours, a consultation with the GI team, and a blood type analysis in case the bleeding continued and the patient needed an urgent blood transfusion. The following day, the GI specialist inserted a scope through her mouth and threaded it all the way down her esophagus to look in the stomach. As the consultant recounted to me later that day, he found in her stomach an ulcer oozing blood, and through the scope, he placed a clip within it to stop the bleeding. When all of the patient's cardiac tests turned out normal, we presumed her chest pain was likely caused by the ulcer rather than a heart attack. Her hemoglobin fell no further, and she was discharged home in stable condition.

I credit that patient's case with opening my eyes to the importance of feces in the practice of medicine. The body's waste is powerful, offering crucial clues about my patients' hidden insides that can help to inform clinical decisions about diagnostics and treatment. And once I had broken the ice with my first rectal exam, I was ready for the next one.

n my first two years of medical school, before feces became a topic of bedside conversation with my patients, I learned about it from an anatomical and physiologic perspective. I studied the gastrointestinal tract's path burrowing through the body from mouth to anus, a remnant of the tubular blueprint that forms in the womb. The two ends of the alimentary canal are two poles of the human body, with the GI tract strung between them forming the axis on which our daily lives spin. As we eat, food snakes through us—entering the mouth, traveling to the stomach, and moving on to the small intestines before traversing the colon. Along the way, the body harvests from it whatever it needs while leaving the rest to pass farther downstream on a peristaltic conveyor belt.

During its journey, food turns into stool, acquiring its typical brown color from the by-product of red blood cell breakdown, and returns to the outside world once again at its terminus. What begins as edible nourishment with pleasing tastes and aromas entering the body through the face—the body's wholesome, front side—metamorphoses into smelly fecal waste exiting the body from its more shameful, hidden posterior side.

The most basic tenet of hygiene teaches that these two poles of the body should never meet, that feces and food should always be kept separate. Human feces are a source of contagion teeming with bacteria and other potentially infectious microbes, and they are the medium through which many infections spread from one person to another. Feces are the epitome of the odious and unwanted, and the orifice through which they are birthed, the anus, is considered the human body's most sacrilegious patch of anatomical real estate.

But studying feces from the textbook perspective of anatomy and physiology never prepared me for what turned out to be a large part of my job—broaching the topic of stool with complete strangers. Besides the grossness and the infectious potential, defecating is among the body's most personal activities. In the privacy of a conversation between doctor and patient, however, a socially unacceptable topic of conversation becomes a necessary subject for in-depth exploration. When it comes to a patient's health, social mores surrounding feces must be checked at the exam room's door.

Getting over both my disgust and awkwardness turned out to be a quick and easy process. Within a few weeks of performing my first rectal exam, I had already conducted several more and had discussed feces in depth with dozens of patients. I no longer needed to avert my eyes when broaching the subject, and my coordination in juggling the lube, the card, and the patient's buttocks improved dramatically. I learned to remove my gloves with finesse once I had finished the exam, pulling one inside out over the other and throwing them both in the garbage in one smooth motion. Though I became increasingly comfortable talking about stool—the social implications having been eclipsed by the medical implications—my patients were not: on more than one occasion, my forthright questioning, together with my nonchalant mention of the need for a rectal exam, was met with flushed cheeks and wide-eyed surprise.

My unabashedly probing questions about a patient's bathroom habits typically hit on various characteristics of feces, including color, frequency, and consistency, and even smell. Many people believe that all feces smell bad, but a cornucopia of intestinal pathology introduced me to far fouler odors than that of healthy, everyday stool. Gut infections like giardia cause a distinct sulfurous odor, while in Crohn's and celiac disease, intestines fail to absorb fat from the stream of food, making greasy stools that float atop toilet bowl water and resist flushing. In these medical conditions, feces can smell like rancid butter.

Stool consistency and frequency were tricky ones. *Constipation* and *diarrhea* are words with amorphous meanings that vary from person to person. Stool is soft and frequent for some, hard and rare for others, and there are no consistent cutoffs for differentiating the normal from the abnormal. One patient might claim to be constipated when stooling every other day, but the same frequency might be normal for another person. Simply asking a patient if they are constipated might be the most useless question in all of clinical medicine. To know for sure, I needed to delve into the messier details of straining and stool shape, and I sometimes used the consistencies of food, such as soft-serve ice cream, for comparison.

The same goes for diarrhea. Early on in my medical education, a senior doctor I highly respected gave me a piece of poignant advice: "To really know if someone has diarrhea," he said, "ask them: 'Are you pissing out your butt?'" His phrasing became my go-to choice of words, and patients suffering from precisely this problem rarely failed to chuckle in recognition.

One morning several weeks into my internal medicine rotation as I was finishing rounds, I heard yelling coming from a patient's room. When I arrived at the door, I saw a nurse dressed in a bright yellow gown standing beside the bed of an elderly man. She was struggling to hold him up on his side with one hand while cleaning the soiled bed beneath him with the other. The man hollered in discomfort. As a fresh, eager-to-please medical student, I walked into the room and volunteered to help despite not being a member of the team responsible for this particular patient. The nurse thanked me effusively.

The patient was suffering from the most virulent type of diarrhea, an infection caused by the bacterium *Clostridium difficile*. I had learned about *C. diff* in class and knew it as a serious, aggressive infection that can be resistant to treatment. Legendary in its infectivity, *C. diff* often sweeps through entire medical wards like wildfire, and in severe cases, patients require surgical removal of the entire colon as the only hope of a cure. The yellow gown the nurse wore over her scrubs was hospital policy—an extra layer of protection donned only when caring for patients with the most infectious microbes. This was the first time I'd seen *C. diff* in a real patient, and the smell was overwhelming. I immediately concluded that volunteering to help had been a grave mistake.

I donned a matching yellow gown and gloves and positioned myself beside the patient's bed opposite the nurse. With one of my hands on the man's shoulder and the other on his hip, I held him up on his side so that she could use both her hands to wipe down the bed and the patient. The man continued to moan, and his body felt weak and limp under my hands, exhausted by uncontrollable diarrhea. The dark brown, watery stool beneath and on him had an intense prickly smell, and I held my nose to the side while struggling to keep a straight face. This episode offered a good but hard lesson in maintaining my professional demeanor even when faced with one of the human body's most horrific smells.

The nurse wiped him off and changed his bed linens, all while wondering aloud why the physician team caring for him had not ordered a rectal tube, a device that would adhere to his bottom and neatly drain all his profuse diarrhea through a tube leading to a large collection bag. She said it would save her the need to constantly clean the poor man, a difficult job for one person. I muttered something about bringing up her idea with the resident in charge.

She looked up at me, mid-wipe. "You're so nice!" she exclaimed, still smiling as her gaze returned to the task at hand. She was middle-aged with highlights of pink in her short brown hair. I held my breath against the smell and hoped she would hurry up and finish.

As she began unfolding a new, clean diaper for the patient, she popped the question: "So, what are you doing after work?" Still holding the man on his side, I turned toward her only to see a coy smile on her face. I stared for a moment in disbelief, still unsure if she was asking me out. She grinned at me: "I was just wondering if you'd like to hang out." My cheeks felt warm and flushed, and all I could think about was whether our patient was hearing this flabbergasting conversation.

I was still new at this, but for her, a more experienced healthcare worker, even the most noxious-smelling feces had faded into the background of clinical medicine. In her mind, she had asked me out over the hospital's equivalent of a water cooler conversation. I declined because I was not single at the time, though the smell of C. *diff* diarrhea didn't help her case.

The following year, between my third and fourth years of medical school, I traveled to India for an independent global health elective. I went for the same reason as many other American medical students: it was both an excuse to travel as well as an opportunity to learn how medicine is practiced in a different cultural context. While there, I worked at a large public hospital in Mumbai with crowded wards and hallways overflowing with destitute patients and their families—a far cry from my medical school's hospital in New Jersey. In India I witnessed my pathology textbook come to life, as I evaluated patients with tuberculosis, typhoid, rheumatic heart disease, leprosy, and polio—all diseases I would likely never see during an entire career spent practicing medicine in the United States.

But my biggest change in perspective on the human body and medicine came with my own bout of diarrhea. For Western travelers in India, infectious diarrhea is an ever-present threat. Many visitors experience it, and those in the country for an extended period can expect several bouts. "Delhi belly" is so pervasive that any shame of discussing diarrhea melts away, and talking about feces with other Westerners became as commonplace as talking about the weather: solid stools are a sunny day, semisolid is partially cloudy, and gas pains are an overcast sky threatening to unleash a thunderstorm (the meteorological equivalent of pissing out your butt). Western travelers in India and healthcare workers are two groups of people wildly open about discussing every dirty detail of defecation. And I was both.

For the first three weeks of my trip, I stood vigilant guard over my bowels. I brushed my teeth only with bottled water, studiously avoided getting any water in my mouth while showering, and ate food only from overly clean restaurants catering to foreigners. The delicious smells from street food carts were tempting, and I jealously watched the Indian medical students living with me in the boys' dormitory devour panipuri, paneer tikka masala, and biryani.

Eventually I had somehow convinced myself that my intestines had adjusted to India, and I drank tap water from the metal cup placed on my table at a restaurant—and that's how I came to discover that Indian food can be as spicy on its way out of the body as it is on the way in.

The day after my show of gastrointestinal false confidence was hot and muggy, but I was shivering cold. An ominous rumbling in my stomach in the morning had flowered into profuse watery diarrhea by midday. I squatted awkwardly over the porcelain hole in the ground in the dormitory bathroom, the only toilet available, and I clutched my knees tightly against my chest, hugging them close for whatever scant warmth they could offer as my body quaked uncontrollably. It was my first time experiencing the chills—I had asked hundreds of patients about this sign of infection, but I had never before experienced it myself.

My body did not easily assume the squatting position that seemed so natural for people in India. My knees bent until I could feel the popping sensation of their ligaments twanging, and my thighs burned as I struggled to keep my heels flat on the ground. My dehydrated and exhausted state further heightened the herculean effort it usually required to stay balanced.

Frequent trips to the bathroom continued into the evening, as large bats mingled with kites in Mumbai's darkening sky. The silver lining was that I had not seen blood or mucus in my stool, which would have been evidence of a more concerning bacterial or amoebic infection. I strolled through the streets around the dormitory where I was staying, stretching my stiff legs, until a sudden urge in my gut sent me scurrying back to the bathroom for the sixth time that day. Both I and my anus were exhausted. Even the soft toilet paper I had sought out at a specialty market had left my backside irritated and bloody, and I could not bear another wipe. In the midst of my agony, a thought that I had up to this point refused to entertain entered my mind: perhaps I should try the Indian method of wiping instead.

I first learned precisely how Indians wipe differently than Westerners on the day I arrived in Mumbai. While being shown the dormitory accommodations, I noticed the conspicuous lack of toilet paper rolls in the bathroom and asked about it—one of the Indian medical students, who went by the nickname KK, was eager to explain. KK was from the southern state of Kerala and wore round John Lennon glasses, and he told me that Indians rub the anus clean with fingers of the left hand while simultaneously pouring a stream of water directly over the area with the right hand. The use of water to clean made the method sound similar to a bidet used in many countries, but this was a manual version with the added detail of direct hand contact with the body's most unholy bit of anatomy. When I asked about hygienic concerns, KK explained that Indians wash their hands with soap and water after performing their ablutions.

He admonished me: "Never ever shake hands or eat food with your left hand." Though I was comfortable with conversations about feces, and rectal exams had become rote, wiping with my hand seemed too radical for me. I resolved never to leave the dormitory without toilet paper.

But now, nearly a month later, the diarrhea and toilet paper had both worn me down. I decided to try wiping the Indian way. In the corner of the bathroom stall was a small faucet coming out of the wall at ankle height, with a small plastic pitcher beneath its spout. I spun the faucet's metal knob and filled the cup with water. Holding it in my right hand, I reached behind me and slowly tilted the pitcher. Water splashed and cascaded down my lower back and careened over the intended spot, while my left hand did what it needed to do.

It was instant enlightenment. The stream of soothing water felt so much better than repetitive assaults with dry, irritating paper. The job was quickly finished, with no need to repeatedly wipe again and again until clean. And thoroughly washing my hands with soap and water afterward eliminated the risk of transmitting my illness to others.

I thought about how intestinal infections spread easily in the United States despite the use of toilet paper, which, I concluded, offers little microbial protection for the hand, but more of a psychological separation from our own anuses. I used water again during my seventh and eighth trips to the bathroom that night, and my coordination improved. The superiority of the Indian method as a quick, thorough, and soothing alternative became obvious, and any shame or hesitation I had melted away.

My diarrhea eventually subsided, but the Indian way of wiping stuck with me. My perception of my own body slowly changed as the Indian perspective on hands became more firmly embedded in my subconscious. Over the following weeks, I became more aware of when and how I used my right and left hands, as if they were yet another two poles of my body that should never meet.

Intestinal infections are among the most common human afflictions throughout the world, partly because diarrhea is a brilliant strategy used by microorganisms for spreading and infecting new human bodies. Just as infections like tuberculosis or coronavirus tickle the lungs into coughing and spread through the air, and sexually transmitted microbes like syphilis and gonorrhea hitch a ride on the human urge to copulate, intestinal infections cause stool to be watery and voluminous in order to get into the water supply or onto a person's hands to spread to their next victim.

While on my pediatrics rotation in medical school in the United States, I cared for many children suffering from diarrhea and dehydration. They received IV fluids for a day or two, and when they could drink by mouth and stay hydrated on their own, they were discharged. Diarrhea was among the most ho-hum pathologies in all of medicine.

In India, I was shocked to learn that the same condition causes children to die by dehydration in large numbers, killing roughly three hundred thousand Indian kids every year. Many of these deaths could be prevented with a clean water supply, simple oral rehydration solutions, and soap for better hand hygiene. The few hospitalized children with diarrhea I cared for in Mumbai were the lucky ones—they made it to the hospital in time, and their families could afford simple IV fluids, a treatment that I take for granted. Diarrhea from a global perspective shows that the major preventive and therapeutic health advances of the twentieth century that have revolutionized life and health in much of the world still have not filtered down to the bulk of humanity.

I realized that the Indian perspective on left and right hands is a method for preventing the spread of these infections, a cultural adaptation reflecting a biological imperative. Such infections are ultimately contracted in only one way: by accidentally ingesting the microbe-laden feces of another person, germ theory's ultimate evil. Anatomically, the body's two hands are symmetric, but through the lens of Indian tradition, they become intensely asymmetric as a simple method of reducing the spread of diarrhea. Still, making the left hand taboo is not all that effective, and neither is toilet paper.

We defecate and clean ourselves afterward not only in a cultural context but also in a socioeconomic one. Traveling in India put on stark display for me the class ramifications that surround how we use the bathroom. I had initially assumed that all Indians' bathroom habits were the same, but later in my trip, I learned that KK and some of the other Indian students could not squat. Like other well-off Indians, they grew up using "Western" toilets instead of squatting over holes in the ground. KK's ankles and knees refused to bend—even mine were more flexible—forcing him to rest on the balls of his feet. The inability to squat was a marker of status and education level, as the positions our bodies assume while stooling reflect a medley of upbringing, culture, and socioeconomics.

Only a very small proportion of humanity has the luxury of access to toilet paper and utilizes it as we do in the States. The Indian method is how much of the world's people clean themselves after defecating, and in other regions, even using water is considered a luxury. A medical colleague of mine who grew up in rural Tanzania told me that, as a child, he could not have imagined using water to wipe himself.

"People would have thought you were completely crazy," he said with a chuckle. In his village, people went into the bush to stool, and afterward used leaves or a rock, or "whatever was available" to clean themselves.

My bout of diarrhea showed me how fortunate I am to defecate in the socioeconomic context that I do, but it also impacted how I practice medicine. I now recommend the squatting position for some of my patients with constipation and its anal fallout, including hemorrhoids and fissures. Squatting's more anatomical position reduces the need for pushing, and even bending the knees slightly beyond a right angle can help stool come out with less physiologic strain. I also instruct the same patients to clean with water to improve hygiene and soothe irritation. Since bidets are not common in the United States, I suggest using a squirt bottle. Judging by my own perception when I arrived in India, I assume people won't be willing to use their hand. I'd be too embarrassed to bring it up, anyway.

Another lasting benefit from my trip to India was that wiping the Indian way means never having to dread the thought of running out of toilet paper. During the coronavirus pandemic of 2020, when there was an unprecedented run on toilet paper, I knew I always had another method to fall back on—one that, in many ways, is even better.

Years after my India trip, while I was working as a hospitalist at MGH, I cared for Paul, a man whose life was being ruined by C. *diff*, the same tenacious infection that a nurse with pink highlights in her hair thought was grounds for a romantic proposition. For Paul, the usual antibiotics had failed three times already, and he was running out of options. I admitted him to the hospital looking haggard, pale, and dehydrated, his face emaciated and his eyes sunken into his face.

For further guidance on how to proceed, I consulted with an infectious disease specialist at our hospital, Dr. Libby Hohmann, who I'd heard was developing an experimental therapy for patients just like Paul. Over the phone she recommended I stop all his antibiotics—the medicines I believed were the only thing standing between him and needing to have his colon removed. She would bring the new therapy once she was certain the antibiotics had cleared out of his system.

Two days later, as I exited another patient's room down the hall, I noticed a petite middle-aged woman wearing sandals standing outside Paul's room.

"Are you Dr. Reisman?" she asked. I nodded as she held up a Styrofoam box in her hands. "Here it is." I had never before seen a medication hand-delivered to a patient's room by a subspecialty physician, which I took to be an indication of the therapy's inestimable value. I followed Hohmann into Paul's room, where she stood at his bedside and removed the box's cover. Inside was a typical orange-tinted medication bottle sitting on a bed of dry ice. And inside the capsules was none other than human feces.

Officially called fecal microbiota transplant (FMT), this novel therapy rests on a very new understanding of the human body's microbiome, the panoply of microorganisms inhabiting every one of the body's crevices and crannies. The human body depends on its internal microbial ecosystem to stay healthy, but antibiotics used to treat an infection often wipe out the good bacteria living in our colons as collateral damage while eliminating the disease-causing strain they are meant to target. Paul was treated with a broad-spectrum antibiotic several months earlier for a respiratory tract infection—reviewing his chart made me suspect it was actually caused by a virus and not bacteria. The antibiotics given by his primary care doctor probably offered no benefit while killing the microbes in his colon and clearing the way for C. *diff* to take up residence.

It makes perfect theoretical sense to restore the decimated microbes by harvesting stool from a healthy person and delivering it into a sick person's colon. Hohmann had me stop all of Paul's antibiotics two days before the FMT treatment so that the healthy microbes inside the capsules would not be killed while traveling through his GI tract from mouth to colon. Though still experimental, FMT has slowly gained acceptance as a treatment for such refractory cases of C. *diff*.

I asked Paul how he felt about ingesting the uningestible.

"I just hope it works," he groaned. Hohmann handed Paul one large capsule at a time and he dutifully swallowed them, fifteen in all. And the following morning he swallowed another fifteen. Within two days, I began to see improvement in his diarrhea, along with a lessening of his cramping abdominal pain. And three days after starting FMT, Paul was looking less pale and sickly and was well enough to be discharged home. Consuming the stool of another person—something many would expect only from a beastly cur—had proved to be his salvation.

I asked Hohmann for a tour of her lab, where her team produces the FMT capsules, and a few weeks later I found myself riding the elevator up one of MGH's many research towers. I followed behind her as we walked the length of a typical laboratory counter that took up most of the room. The tools of her research were air-drying at the far end—a blender and a few steel strainers. Hohmann pointed to a large white freezer nearby, where she kept the stool, but unfortunately it was empty of specimens the day that I visited. To my surprise, the lab didn't smell at all like a public restroom.

As we stood beside the typical kitchen appliances, Hohmann explained how she prepares the capsules. Frozen stool is thawed out the night before the therapy is administered, and in the morning, she blends it with an added splash of salt water. She strains the resulting slurry through the large steel sieves on the table to remove debris and solid chunks, and injects the final product into gelatin capsules that patients can swallow. It sounded like a recipe for a rich French dessert.

When we walked into Hohmann's office and sat down to chat, she passed me a Styrofoam box identical to the one she'd brought the morning Paul received his treatment. Inside it were three medicine bottles, and I could see the capsules inside, each filled with a small smudge of brown. Hohmann admitted that see-through capsules are not ideal, since patients can actually see the stool inside them. Opaque ones, however, are much more costly, and her research budget did not allow her to spend extra money on this detail.

Transparent capsules also came with another challenge: her study needed an authentic look-alike for human stool. As with any drug trial, one group of patients must receive an inactive placebo against which to compare the experimental treatment's effects. After some experimentation with a variety of gravies and chocolates, she settled on a mixture of cocoa powder and gelatin. But the capsules in the Styrofoam box in my hands were, she insisted, the real thing.

Because the therapy is so new and experimental, there was no way for Hohmann to source the stool, except from the humans who produced it. Potential donors are extensively tested for a variety of infections, including most of the diarrheal diseases common in India and elsewhere. In Hohmann's words, donors should be "screamingly healthy" and cannot have even traveled to countries like India for over a year. It had been two years since my trip, and I asked if I could donate—as it turns out, no physicians or other healthcare workers are allowed, because of their fairly regular exposure to C. *diff*. Wise, I thought. Hohmann also tells donors not to eat peanuts in the days before donating, just in case a recipient of the transplant is allergic.

She rattled off a few statistics from her research—in refractory C. *diff* infections, the usual antibiotics cure less than half of people. She found much higher cure rates with FMT—87 percent after the first two-day course, and 92 percent after a second course. Despite the success, however, Hohmann's research suffers from bad public relations. Many administrators at MGH, when first informed of her proposed research project, found it "too gross for words." Patients are also reluctant to participate—some have told her they'd "rather die" than take FMT. And some probably did.

Hohmann hopes to expand her study to include patients with their very first occurrence of C. *diff*, rather than just relapses, but she has been barely able to convince anyone at all to participate. She discovered that patients warm up to the idea of taking FMT instead of the usual antibiotics only after repeated bouts of severe diarrhea truly start to ruin their lives. I thought back to my time in India and was not surprised to hear that severe diarrhea has a unique ability to open a person's eyes to a new perspective on the body and its effluents.

"Which therapy would you prefer if *you* got C. *diff*?" I asked her, playing the physician's version of the game Would You Rather. She responded immediately: FMT. "Me too," I voiced without hesitation. Perhaps healthcare workers like Hohmann and me approach this question differently than the general public. We have been through similar training and were well aware of the risks of everyday therapies like antibiotics. Also, with our profession necessarily comes an unusual comfort with the stool of other people.

Medications have always come from the natural world, from the bodies of animals, plants, and fungi, and the human body is merely another corner of nature from which medical treatments can be extracted and refined. Human blood is a lifesaving therapy in transfusion; the antibodies of another person's serum can help those suffering from infections like coronavirus, hepatitis, and rabies; and human flesh in the form of donated organs remains the only treatment for many kinds of end-stage organ failure. Now, in addition to our flesh and blood, human feces have joined the list of lifesavers harvested from our own bodies.

On its face, FMT sounds like the worst advice a doctor could ever give to a patient, a therapy that flies in the face of every hygienic and health principle that medical practitioners like me have cherished. While accidentally ingesting another person's feces is what kills hundreds of thousands of people throughout the world every year, we now know that the same revolting act can also save lives. Based on the success of FMT in C. *diff* infections, it is now being tested as a therapy for a wide variety of other medical conditions, even ones beyond the gastrointestinal tract. But for it to work, people will have to change their perspective.

Feces are a physiologic fact, as is the anus, a part of our bodies that we can never directly see. But our perception of our own bodies, even their most unholy parts and processes, can change as it did for me during my medical training and while traveling in India. In the future, both physicians and patients may need an even more expanded understanding of our relationship with feces in order to stay healthy and treat disease. Perhaps dogs and other animals that eat feces are not brutish and barbarous, as most of us would presume, but instead engaging in a healthful practice of sharing microbiomes, a strategy from which humans might learn a lot—but only if we can get past what our society has conditioned us to see as gross.

4

GENITALS

When I was a pediatrics resident, I spent a lot of time standing around in hospital delivery rooms staring at the vaginas of women I had never met before. As the women pushed their hardest for those last few exasperating inches, the baby's crowning head dilly-dallying its way toward the light, I stood to the side and waited nervously. A baby is technically an obstetrician's patient up until the moment of birth, but after that, I would be responsible for the newborn's care. I kept my eyes on the labial curtains through which my patient would make an entrance, and I mentally prepared to make mine.

Birth entails some of the most exciting and nerve-racking moments the human body has to offer, and some of the tensest moments in all of medicine. I was usually summoned to the delivery room when a warning sign worried the obstetrician about the unborn baby's condition. Sometimes it was the flow of cloudy fluid when the woman's water broke, a sign the baby had stooled inside the womb; other times the reason was an abnormally slow rhythm on the fetal heartbeat monitor wrapped around the laboring mother's belly. When my pager beeped with a room number and a reason, I rushed so that I would be present and ready for the moment the baby exited its mother's body.

Newborn babies are ultimately products of our genitals—specialized organs designed to make new human bodies. The single moment of birth is the culmination of nine or so months of a fetus growing and ripening like a fruit within its mother's genital tract, and my immediate task after delivery—and the main reason I was paged by the obstetrician—was to ensure that the baby started breathing. At birth, every baby enters a completely new world with an atmosphere, and my job was to help newborns that struggled through this colossal transition. In the large majority of cases, even when the warning signs seemed dire, the baby was born and cried immediately, its rhythmic wailing a sign that the child was inhaling and exhaling appropriately. That sound told me that my job in the delivery room had ended before it began. Whenever I heard crying, I pulled off my still-clean gloves, congratulated the new parents, and left the room without having touched the baby or done a single thing other than watched and waited.

During one busy overnight shift, newborns were popping like popcorn— I received page after page and stared at vagina after vagina. In each and every case, I never knew for sure what would happen once the baby finally came out: Would it breathe on its own, or would it fail to breathe and need my help?

One newborn's birth was greeted by an ominous silence. The obstetrician delivering the baby quickly clamped and cut its umbilical cord, and passed me the lifeless bundle. The baby's arms and legs felt limp like a rag doll's, and once I had the baby under the warming light, I saw the intense blue tint highlighting its underlying brown complexion. The baby's chest was completely still with no respiratory effort at all, and I knew I had about sixty seconds to get breathing started for the baby to have its best chance of surviving.

I grabbed a towel and rubbed it along the baby's spine with my fingers in an attempt to jump-start breathing. Usually the mild stimulation provided by this maneuver is enough to activate a newborn's lungs. In this case, it did nothing. I then reached for a rubber mask, put it over the infant's nose and mouth, and squeezed the attached bag, nudging a puny puff of air into its tiny lungs. When I squeezed and released the bag a second time, the baby finally started moving. Its limbs curled with the measliest signs of life, and I heard a faint cry through the rubber mask still over the baby's face. I squeezed the bag a third time for good measure and then pulled the mask away as the baby began wailing with the vigor of life.

The rhythm of breathing had begun in earnest, and it quickly turned the tint of the newborn's skin from blue to pink. Pink meant that this baby's lungs were doing their job—that precious air was getting where it needed to go inside a new body. Getting a newborn to breathe feels like starting a lawn mower—a single tug on the rip cord initiates the beat, and from there a baby breathes on its own, whirring into the future in a self-sustaining respiratory rhythm that would persist for the rest of its life.

Once I felt confident that the baby had gotten the hang of breathing, I congratulated the new parents, told them how vigorous and healthy their new baby was, and left the room to go write a note in the medical chart about the birth—a story of starting the beat.

Each newly minted human body means the initiation of many new rhythms beyond just that of lungs inhaling and exhaling. A baby's heart begins its own rhythmic beat months before birth, and I often measure it by pinching the gelatinous remnant of a newborn's umbilical cord between my thumb and forefinger—the blood vessels coursing through the umbilical cord seal themselves off within a few days after birth, but they briefly offer an easy way to measure a newborn's pulse. In the days after birth, other rhythms of the body join in as well—sleeping and waking, eating and stooling, crying and soothing.

We are clockwork creatures, and almost everything I learned about the human body as a medical student was a rhythm. In adults, the heart beats about once per second, keeping roughly the same time as the second hand circling over an analog clock face, and lungs inhale and exhale with the same cadence as ocean waves pounding the shore and retreating. These are the body's most fundamental rhythms, which is why every patient visiting the doctor has them checked as part of the "vital signs." A cuff on the upper arm measures blood pressure, while the musical drumbeats of heart and lungs are appraised as the bedrock of human health.

But there are many more beats in the body, each an actual biological clock keeping its own idiosyncratic time. Eyes blink every few seconds, and blood cells circle through vasculature every five minutes or so. Regular meals lead to (hopefully) regular excretions, bladders and intestines filling and emptying in dependable quotidian replay. For many humans, the ingestion of intoxicants also follows a regular cycle—a cigarette brings a rush of nicotine every hour, a shot of heroin several times per day, or a regular drink of alcohol every evening. The human body depends on the rhythms of habit and can get used to, and become dependent on, almost anything— even the drumbeat of inebriation.

Though our lives seem to proceed in a linear fashion with one day following another and days progressing inexorably forward, human biology is best understood as not a single consecutive progression but a complex circular weaving of overlapping and interlaced musical meters. Homeostasis is an intricate blending together of beats, with melodies varying in tempo as they go round and round.

The strangest rhythm in the human body, and the one that seems to break all the rules, pertains to genitals. Menstruation is the drumbeat of the female genital tract, and it entails the uterus shedding its inner lining, a uniquely transient part of human anatomy that repeatedly comes and goes with a calendar's regularity. For this rhythm, the brain's pituitary gland serves as the conductor, secreting into the bloodstream hormones that coax the ovaries and uterus as dutiful musicians through their cycle.

Menstruation may be the body's slowest rhythm, beating only once a month, but it is also one of the tardiest: over a decade usually passes after birth before it finally kicks in. Male genital tracts have their own rhythm—a daily cycle of testosterone secreted by the testicles—and it, too, begins its beat around the same time during the teenage years. Puberty is the human body's springtime, when genital tracts bloom like flowers and begin offering new and unrivaled bodily pleasures like sex and also masturbation—a completely natural act in which the body's dexterous hands and fingers trick our genitals with a semblance of sex. Our genitals are extremely gullible, a convenient discovery for many going through puberty.

Menstruation is a bodily rhythm much more widely recognized than that of the male genital tract, and a far more mythologized one. Because of its monthly periodicity, many have theorized that menstruation is synchronized with the moon's phases, which is why the words *menses, moon,* and *month* all share a word root. The uterus's lining acts like the body's own lunar calendar, growing each month like a sliver of new moon waxing into a gibbous convexity before shedding again. Studies of the moon's phases have shown only inconsistent alignment of the two rhythms, and only in the most rhythmic female genital tracts. The purported menstrual synchrony of women living in close quarters also appears to be an inconsistent finding.

The rhythms of our genitals are unlike all others found in the human body in another crucial way: they are not necessary for the body's survival. Every other tempo contributes something to the rhythmic foundation of health, but not the genitals. In this way, they are like flowers—specialized organs used by plants to make more of themselves, but ones that do not directly serve the plant to which they are attached. Genitals are, of course, necessary for the survival of the human species, but not for the health of each human being. They are uniquely future-oriented body parts, programmed only to promise parenthood at some later date.

Though genitals are not truly necessary for the biological survival of each human body, they make life much better, and this is something they share with music. Like many people, I became more aware of my own genitals as a teenager around the same time that I became deeply cognizant of music. During this volatile age, hormone levels in the body burst and genitals mature, and for some of us, music has a profound emotional impact on our lives.

It was around this age that I first learned to play guitar, and I regularly got together with friends to play music. We never formed an actual band, but I learned to stay on beat and began to understand how multiple instruments can harmonize with one another. I trained my tapping foot to feel and match rhythms, and trained my ear to recognize the confluence of chords and melodies. Music, I realized, is an interwoven assemblage that adds up to something greater than the sum of its individual parts. For many musicians, playing an instrument or being in a band also leads to an initiation into sex and the genital world of rewards and regrets, but I was not so fortunate until well after high school ended. I did, however, develop a lifelong appreciation for and understanding of rhythm and music, and I carried those with me into my career as a physician.

In medical school I realized that feeling the human body's rhythms was an important part of differentiating healthy states from disease. I learned to place two fingers against a patient's wrist or neck to feel for a pulse and count the tempo; and I counted a patient's respiratory rate in the same way—this was more difficult since people have partial control over their own breathing, and drawing attention to it often causes its rhythm to change. Being a student of medicine meant being a student of the body's music, and it required becoming intimately familiar with rhythms and learning to size them up in an instant.

When all of a patient's rhythms were within a normal range, it meant they were healthy, but I learned to recognize when the body's rhythms faltered in disease. When feeling for a pulse in certain patients, I found their heart beating dangerously fast, thumping like techno music at a rave; in others, it was beating too slowly, necessitating the emergent insertion of a pacemaker as an implantable metronome. Sometimes a patient's heart rhythm was totally off-kilter, beating with the capricious syncopations of a jazz drummer instead of a steady beat. Similarly, when the rhythm of breathing slowed too much, a person might need to be put on a ventilator, while breathing too quickly was an important indication of respiratory distress.

I assessed all of my patients' rhythms in a similar way: stooling can speed up into diarrhea or slow down into constipation; a sped-up tempo of urination can be a sign of diabetes or urinary tract infection, while a slowdown could mean dehydration or a swollen prostate blocking urine's flow. Most people would easily recognize when a familiar song changes tempo, and I came to know the rhythms of the human body just as intimately and could recognize when they were off. And if one organ system loses its proper beat, it can throw off the body's greater harmony.

But I was trained to become most alarmed when bodily rhythms ceased altogether. Rhythm is life, and when it stops—when the heart stops beating or lungs stop breathing—certain death usually follows. But genitals break this rule too. When sperm and egg manage to find each other, they fuse and begin to grow into a fetus, and the regular beat of the female genital tract stops. For menstruation, an abrupt pause is not a case of organ failure at all—rather, it is a case of organ success. Unlike everywhere else in the human body, the interruption of this rhythm is, in fact, the whole biological point of the rhythm in the first place. And when that beat halts, a human body steps outside of the usual repetitive cycle of life and heads off on a complete physiologic tangent—pregnancy.

My own journey toward procreation began the day I met Anna, the woman I would marry. It was at a meeting of the Explorers Club, and we connected as lovers of adventurous travel to the world's geographic extremes before we became lovers of each other.

Soon after meeting, we took a trip abroad together to Serbia, where we attended the "Test Fest," also known as the World Testicle Cooking Championship. It took place in a nondescript field in a rural part of the country, and for the past four years, my new girlfriend had been the reigning judge. This bizarre cookout combined innumerable varieties of testicle goulash with staggering amounts of alcohol. The men in attendance insisted that eating testicles increases virility, and they were so inebriated that I did not bother to point out that testosterone is completely metabolized in the liver after being ingested, so none of it actually reaches the bloodstream. This is why testosterone supplements come in the form of a shot or a foam to be absorbed through the skin; with female hormones, the opposite is true—estrogen and progesterone stay intact when taken by mouth, hence "the pill," an ingestible method of tricking the gullible ovaries into withholding each month's egg. Many of the Serbs would have been disappointed to learn this fact, though probably not so disappointed as they appeared to be that, this year and for the first time, Anna showed up with a partner.

It was some years later that Anna and I were on an adventure in the mountains of another country, when her usually regular period didn't come. When the monthly beat halted, we knew: she was pregnant.

Before I watched my own wife on this unique anatomic journey, I had cared for pregnant women during my obstetrics and gynecology (OB-GYN) rotation as a medical student, and I found pregnancy to be by far the body's most unusual process. Unlike the rhythmic mechanisms everywhere else, pregnancy is one-way only. Instead of pulsing repetitively with a beat, it slowly builds in intensity and eventually peaks in an explosive climax in which a baby is launched from the body into the outside world. In caring for my pregnant patients, I measured the dome of their growing bellies with a paper ruler, a visible manifestation of pregnancy's swell toward crescendo.

With other medical ailments, I learned to offer treatments in hopes of returning a patient's body back to the way it was before. Whether in a hospital ward or an outpatient clinic, my goal was to make a rash go away, to help symptoms resolve, or to restore organ function after a physiologic insult knocked things askew. But pregnancy's inexorable march toward birth's eruption turned the usual clinical consideration on its head.

Everything about pregnancy seemed oddly explosive, including its beginnings in another kind of biologically climactic process. Sex is what lights the fuse that leads to birth—it starts out rhythmic but builds in intensity, a positive feedback loop of pleasure and need egging each other on. The process crests in orgasm with the launch of semen, a projectile glop made of sperm and its food for the trip.

Though orgasm depends on the unique sensitivity of our genitals, it takes place as much in the brain and adrenal glands as it does in structures of the pelvis. Peaking orgasm comes with a surge of adrenaline—the heart pounds, and the brain is fixated and momentarily utterly useless. And after climax, the wave of orgasm passes quickly. The adrenaline surge recedes like a spent tide, and heart rate and blood pressure return to normal. The pituitary gland sprinkles into the bloodstream hormones like prolactin and oxytocin, which bring a postcoital calm to replace what was a thunderous and unquenchable urge just moments before. Here physiology does something that it almost never does—it switches on a dime. And after the explosion, nothing seems quite the same anymore.

A similar launching takes place in the female genital tract when the egg is released from the ovaries in ovulation—a surge of hormones sparks the detonation, and a microscopic ovum is sent hurtling through the fallopian tube and toward the uterus. Like sperm, the egg is sent flying into an uncertain world like pollen released by flowers in hopes of finding a mate. As a medical student studying the physiology of it all, I learned about how many things can go wrong, and I marveled that the sperm and egg ever find each other at all in that vast darkness. Of course, the large number of babies born every day—especially during busy overnight pediatrics shifts—told me that humans have enough sex to easily flog a biological improbability into a near certainty.

When the two specks of genital pollen do manage to meet and combine, the result is a gestating fetus growing inside its mother's body, just as the pollination of flowers leads to fruit ripening on—but offering no direct biological benefit to—its parent plant. During my wife's pregnancy, I was a bystander who could only marvel at all of pregnancy's bizarre symptoms and discomforts. Each trimester brought new and unique pains and bodily limitations. She lost the ability to enjoy certain foods she used to love, and eventually lost the ability to bend over at all.

After almost nine months, the rhythmic contractions of my wife's uterus gave the first sign that the climax was coming. We counted its tempo care-fully as the contractions fluctuated over days, and when the rhythm became regular and intense, we knew it meant that it was time to go to the hospital. Once pregnancy nears its end and the cascading crescendo of labor begins, there is no turning back—a baby is exiting its mother's body one way or another.

Biology is rhythm, but even when the body steps outside of its usual rhythms—when menstruation pauses to allow for the initiation of pregnancy—it is part of the larger cycle of our genitals, a cycle of reproduc-tion and perpetuation of the species. Each human body, and all its internal cadences, are contained with the larger cycle of the next generation follow-ing the previous one, each another beat in a vast rhythm that has played out since the beginning of life on earth. Though genitals are the great disrupters of the body's usual rhythms, they are the organs that ultimately keep the music going. With the birth of a child, nothing is ever the same, yet human biology continues as it always has in the grand circular sweep of time, and nothing at all has actually changed.

I have been present at innumerable births, and most of them have blurred together in my memory into a swirl of waiting, followed by a shower of blood and a crying infant. But the birth of my own child stood out—instead of arriving only for birth's climactic explosion, I helped my wife through four agonizing days of labor at home before her cervix had dilated enough for her to be admitted to the hospital.

Of all the times I waited around impatiently for a baby to exit a vagina, the birth of my own son took by far the longest. As labor dragged on and my wife pushed through the discomfort and pain, I wondered if our child would breathe right away. When he finally came out, he cried immediately and kept on wailing. At that moment he was—like every newborn baby—a seed tossed into the world like dandelion fluff on the wind with a hope of landing in suitable soil, taking root, and being sustained. My son's cry meant that his body had found its fertile soil—an atmosphere in which his lungs had taken root—and his body would persist into the future, even without the help of a pediatrician like me.

And for the first time, once a newborn baby's rhythm was established, I did not leave the room. This time I stayed, and watched my child in the arms of his mother. And I knew that for us, nothing would ever be the same.

5

LIVER

first encountered the human liver in anatomy lab as a medical student. It was purple-brown and glistening, enthroned in the upper right-hand corner of my cadaver's abdomen just below the rib cage. The liver seemed to lord over all the other digestive organs in its purview. Peeking out from beneath its fleshy edge was the liver's handmaiden, the gallbladder, a storage vessel for bile produced by the liver itself to help digest and absorb nutrients from the intestines.

Over the following months, I got to know the liver in staggering detail. I memorized all its anatomical relationships: the right kidney behind it, the intestines coiled at its foot, and the arteries and veins tying it to the rest of the body's vascular system. I saw zoomed-in views of the liver's many structural facets in a class on histology—the study of tissue on the microscopic level. In biochemistry class, I studied the liver's elaborate and hegemonic control over the body's metabolism and many of its intricate homeostatic balances. The liver also acts as the body's gatekeeper: every bit of food we eat and absorb through our intestines goes through this organ, a checkpoint where nutrients are sorted, packaged, metabolized, and distributed to all of the body's corners.

Like many other parts of the body, the liver became a familiar acquaintance about which I thought I knew almost everything. I could mentally deconstruct it into its parts and components, and reconstruct it back into the whole functioning organ. But it still remained abstract. Even after I had memorized all the necessary information about the liver, my journey of getting to know this organ would continue and eventually encompass the entire spectrum from the cells that make up a liver to the whole bodies of my patients harboring livers within their abdomens. In the process, it would change my perspective on disease, life, and, surprisingly, food.

Once I started working in the hospital and treating patients with liver disease, I was able to put faces like Juan's onto an impersonal organ. And Juan's face had liver failure written all over it. His eyes glowed with the brilliant yellow of a highlighter marker, and his features were cadaverous. The rest of his body told the same story: his abdomen was distended and full of fluid, and his dull yellow skin was a geographic patchwork of purple splotchy bruises. The yellow color, known as jaundice, resulted from his liver no longer being able to clear the bilirubin slag left over as his body recycled spent blood cells. The fluid in his abdomen was caused by his liver failing to synthesize the usual bloodstream proteins and also clogging up its own blood flow. His bruises stemmed from his liver giving up on its usual contribution to helping blood clot. When the liver fails, virtually everything goes wrong.

During residency, I worked in a hospital with an active liver transplant service, which meant that I cared for many patients like Juan, whose livers were in their final throes. In most cases, my patients' livers had been degraded over years by a slow trickle of damage, usually wreaked by alcohol. In a few, liver breakdown was sudden and catastrophic, like the young man who dropped to the ground while running a marathon on a hot day—his liver was fried by the hepatic blitzkrieg common in severe heatstroke. Or the young girl in her twenties who tried to commit suicide by overdosing on Tylenol, a medication that, if taken in excess, can destroy the liver precisely because it is the organ that detoxifies the bloodstream.

Juan's was also a case of acute organ failure: his liver suddenly went into an unexpected death spiral after he took a common antibiotic prescribed by his primary care physician. Though certain antibiotics can sometimes cause mild liver inflammation, Juan's all-out liver failure was an extremely rare and very unlucky side effect, and there was no antidote. He was in his forties, and he had never taken a sip of alcohol in his life. When I started a new rotation through the medical ward, he had already been in the hospital for weeks— yellow, swollen, and too weak to stand on his own. According to the outgoing resident signing him out to me, Juan was high on the liver transplant list, and my job was simply to keep him alive long enough for him to receive a new one.

Over the next month, I monitored Juan each day for the expected complications of liver disease. I had seen such patients bleed to death, and Juan's easily bruised skin reminded me that I had to monitor him extremely closely with frequent blood tests. One day he vomited up what looked like coffee grounds—blood mixed with stomach acid takes on this appearance, and it indicates a GI bleed. I ordered a stat IV antacid medication, a consult to the GI service, and ever more tests and monitoring.

The fluid in his belly stretched and strained at his organs and pelvic ligaments, causing him near constant agony. I ordered him pain medications, albeit in smaller doses than usual, knowing that his dysfunctional liver could not properly metabolize most drugs. When the pain became unbearable, I drained fluid from his abdomen by placing a large needle right through the skin, each time filling several large glass bottles with many liters of what looked like melted butter, foamy on top. The procedure lessened Juan's pain slightly, but his belly immediately began to fill again with fluid. Everything I did for him was a temporary fix until he could get his transplant.

The weeks proceeded with one bump in the road after another, each a testament to the wide-ranging fallout for the body when its largest internal organ quits. I treated each complication as it reared its head and hoped each day for good news from the transplant team. Juan was by far the sickest patient under my care and required the most careful attention. On many mornings during my rounds, I was surprised to find that he was still alive, that he had not died overnight from some complication that had arisen the day before. I marveled that the steadily worsening results of his blood tests were still compatible with life.

Caring for Juan felt like running around putting out small fires even as the entire house was burning to the ground. I felt certain he would die waiting for a liver.

Throughout the month-long rotation, I got no word that a liver was coming, not even a false alarm. Most patients wait months, I reminded myself, though livers usually come more quickly for patients with acute failure, like Juan. As I prepared to move to a different ward, he was sicker and closer to death than when I'd first met him, weaker and with even less of an already nonexistent appetite. On my last day on the ward, I said goodbye to Juan and to his wife, Anna, who had been at his bedside nearly every day. My path in life had briefly converged with both of theirs because of Juan's liver, and I felt a heavy sadness after having been through so many ups and downs with them. I thought of all the times I had been called to Juan's bedside for some new emergency: in each instance, I hurriedly examined him, assessed the situation, and devised a plan as Anna looked on—worry in her eyes, yellow in his. I worried with them, and we had hoped in unison for a transplant. Their medical odyssey would continue, and I caught only a glimpse of its trajectory before moving on. A rotating resident usually gathers little emotional moss, but I felt as though I had formed a special bond with Juan and Anna, and I ached for his survival.

During sign-out—when I handed over care of all my patients to the next oncoming resident—I warned her that I thought Juan was near the end. Still, I told her to keep him alive until he received a transplant. I began working on a different ward, and a new roster of patients pushed Juan deep into my memory.

I had actually met the liver years before I started medical school. When I was a child, chopped liver was a holiday staple in my house, a beloved spread everyone enjoyed on occasions both religious and secular. I alone found it utterly revolting. The drab beige color highlighted by a repellent pink, the granular texture, and the metallic tang of putrefied iron—it simply was not something I ate.

My disgust had nothing to do with the liver's being an internal organ. I never even thought about what liver really was, and never considered its origin inside some animal's abdomen before it ended up sizzling beside onions in my mother's frying pan. I also felt nothing for the creatures that had once depended on those edible livers to stay alive. Instead, my revulsion was a thoughtless, programmed reaction to the sight, smell, and taste of something I considered to be conclusively and immutably "gross."

But once I got to know the liver in medical school, my perspective changed. When I went home for Thanksgiving dinner in the second year of medical school, a bowl of chopped liver sat in its usual place on the table beside slices of turkey meat and cranberry sauce. As usual, I watched my gleeful relatives lavishly spreading their blessed mayonnaise-infused muck over crackers, and the revulsion bubbled up in my throat.

But this time, alongside it, a flood of the liver's detailed biology rushed through my mind. After studying the liver for over a year, my memory was saturated with information about the organ, fattened like a goose's liver bound for foie gras. Looking at a bowl of chopped liver while thinking about all of the organ's essential and life-sustaining tasks in the body—as well as in the bodies of animals whose lives ended to make the dish—made me suddenly see it in a different light. The thought of all that complex biology being reduced to mere food seemed an almost magical transformation. For the first time, I felt an urge to try it.

I smeared a spoonful of beige mush across a cracker and brought it to my mouth, as memorized diagrams of the human liver's glucose production and coagulation proteins swirled through my thoughts. Fighting through my ingrained disgust and the liver's dog-food smell, I shoved the cracker into my mouth. The taste was a rush of minerals and meat, and it was not so offensive as I had remembered. I swallowed it without gagging. And then I reached for another cracker. And another. It was certainly not love, but the spread's connection to what I knew of living livers enchanted me, and my medical education proved to be a catalyst for my rejoining the rest of my family in appreciating a traditional food.

Learning to enjoy chopped liver gave me a new perspective on food preferences in general. Every individual person has a unique definition of edibility and a different perspective on what's gross. More than anything, where someone draws the line between edible and gross is simply a matter of what they are used to eating. Still, for many people, the worlds of anatomy and appetite must remain utterly severed from each other if the latter is to survive at all.

I used to feel the same way. In childhood, I knew meat only as abstracted, disembodied red slices swaddled between Styrofoam and plastic wrap on supermarket shelves. In my naïve mind, all food came from the grocery store, and I had little concept of how it got there. Protected by my own narrow awareness, I was blissfully ignorant of meat's sometimes gory provenance. And I still completely understand why meat's animal origins need to be kept out of mind for some people in order to enjoy a meal.

But once I learned about the body in exhausting detail, even the standard fare of eating animal muscles became more interesting. I cross-referenced cuts of beef against the muscle groups I was learning about on the human body and found that cattle and humans have pretty similar anatomy after all—the muscles just have different names. Cattle have the filet mignon, and we have the psoas major muscle. They have the rib eye, and we have the erector spinae. The same muscles responsible for every movement our bodies make in life suddenly become meat after death, a change in perspective with no actual biological difference.

My own new culinary curiosity did not stop at the holiday dinner table. My liver flirtation spurred a growing anatomical appetite that ventured into other parts of the medical corpus, and my medical and culinary educations proceeded in step. After studying kidneys in nephrology class, I tried steak-and-kidney pie at a local restaurant and discovered that the millions of microscopic blood filters—called glomeruli—that fill the organ's flesh make it substantial and pleasantly chewy. I learned about the pancreas's role in the body's endocrine system and sampled pancreas sweetbreads for the first time. Though the internal organ is typically discarded as offal, I found it flavorful and filling.

In an immunology lecture, I marveled at the staggering fact that hundreds of billions of red and white blood cells pour from a person's bone marrow into the bloodstream every day, and when I first tasted roasted marrow bones, I marveled at a staggering new level of deliciousness. No wonder my immigrant grandmother had been so fond of it. Anatomically, bone marrow is the body's most perfect food for cooking—the cavities within our largest bones are filled with pockets of stem cells couched in fat for frying, and all of it contained within a bone cooking vessel that enhances the flavor. Simply add heat.

Though liver is the internal organ most widely recognized as food, both by people who partake and those who do not, it has the strongest taste of them all. I regret that liver's intense flavor probably turns many people off to all other anatomical eats, though there are numerous tastier options. Because of liver's potent and overpowering taste, it does not make a good gateway organ.

Studying medicine broadened my culinary horizons and showed me that knowledge of anatomy and physiology, mixed with awe at the body's physiologic and anatomical harmony, was a consummate flavoring. Besides, as I realized in medical school, every single part of an animal's body is edible so long as it can be chewed without fracturing a tooth and swallowed without stabbing you in the esophagus. And even those inedible parts? They make a good broth.

The most anatomical meal of my life came during a trip to Iceland toward the end of residency. While visiting a friend outside of Reykjavík, I was served *svid*, a traditional Icelandic dish. I was handed a dinner plate, and staring up from it was a sheep's head split down the middle from front to back. The exposed facial cross section was an image straight from my anatomy textbook: for weeks I had studied that same diagram (though human rather than ovine) leading up to a medical school exam. I recognized the same eyeball sitting framed by orbital adipose and wrapped with miniature eye-swiveling muscles. The serpentine nasal passages lined by paper-thin swirls of bone exposed on my plate were the same ones I had navigated again and again in my mind, learning each detour to the middle ears and sinuses. And the animal's muscular tongue flung against a row of teeth reminded me of my delight at first learning from my cadaver in medical school how the tongue is attached in the throat—a mystery I had pondered since childhood.

My comfort eating animal livers had not quite prepared me for having a face on my plate looking back up at me. The face is far more personal than abstracted cuts of meat or hidden abdominal organs like the liver—we see ourselves as inhabiting the real estate just behind our faces, and the face is how we recognize our friends and loved ones. The sheep's face pointedly invoked the life that the animal had lived and bluntly told me what it had taken to make that meal. With my fork I poked at the pearl-colored stump of optic nerve protruding from behind the sheep's eyeball, the olfactory nasal mucosa, and the taste buds cobblestoning its tongue—these sensory organs and the nerves wiring them to the brain (which is typically removed in svid) had captured a lifetime of sights, smells, and tastes for that sheep.

With the dinner plate as a mirror, I thought about what lay behind my own face. My senses and their circuit board of perception had captured almost every experience of my life, including the sensations of every bite of food I've ever eaten. The svid suggested that I, too, despite a lifetime of experiences, would one day end up on the dinner plate of some creature, be it large or microscopic, everything I touched and saw reduced to little more than chopped liver. Considered from the rigid science of anatomy and physiology, the death through which a living sheep turned into Icelandic dinner was no different from the death I would spend my medical career fighting against, or easing, in my patients. And my own death would one day also be part of the very same process. Just as dissecting a cadaver told me about what my own body was made of and gave me an indirect glimpse inside myself, digging into the most philosophical meal

I have ever eaten told me in no uncertain terms what my own body was made of: food.

With empathy added to taste, I savored every bit of the svid. Its eyeball fat was velvety and luscious, while the muscles of mastication strapped to its jaw, so impossibly tender, were soft enough to be slurped. Reducing life's beautiful complexity to food, like reducing the emotion of love to a mere trick of the reproductive drive, seemed no reduction at all. It was merely a shift in viewpoint.

When my perspective on eating liver flipped, nothing objectively changed. Liver's smell and taste stayed the same, as did my own taste buds, sense of smell, and neural wiring linking sensory organs to my brain. Outwardly, everything was precisely the same as before, except for an imperceptible change deep in the part of my brain where perception is interpreted, labeled, and linked to emotion. My interest in trying liver again was sparked by my own awareness of what takes place behind the meal—concealed within every bowl of chopped liver ever to grace a dinner table is the life of one or more animals, and a complex biological tale of what their livers accomplished before ending up smeared over crackers.

For farmers market shoppers and other proponents of local food, an understanding of who grew their food and where on the land it was cultivated provides a stronger connection to the food that they eat. I have found this to be equally true of foods from animals. Learning how the bodies of humans and animals are constructed and how they function showed me where and how my food once served in life, and it forged a deeper appreciation for what goes into my mouth and becomes incorporated into my own flesh.

Medical school broadened more than my understanding of basic anatomy—it also taught me that knowing good food and how to cook it well means knowing the anatomy and physiology of animals. And with svid bringing the worlds of anatomy and appetite into titanic intimacy, I was able to put a face onto abstract food. To my surprise, food became more personal, and I discovered that empathizing with it and eating it are not mutually exclusive. As for my own sense of what is gross, medical school rewired my habits of edibility and cleansed the doors of my culinary perception. And, as William Blake might have said, when the doors of perception are cleansed, one may see things as they truly are—delicious.

Six months of residency passed as I bounced around the hospital from one rotation to another, and while working in another medical ward, I cared for a young man hospitalized with an intestinal infection causing profuse diarrhea and dehydration. As I hurried through my morning rounds, the young man's nurse called out to me across the ward.

"His father says he knows you!"

I looked again at the patient's name as I rushed to cross things off my long to-do list, but I didn't recognize it.

That afternoon, I returned to the young man's room to discuss the possibility of discharging him home; his condition had improved, and he no longer needed intravenous fluids to stay hydrated. Next to his bed stood a skinny man in jeans and a baseball cap, a surgical mask covering his nose and mouth. Before I could broach the subject of discharge with my patient, the man turned toward me and pulled down the mask to reveal his face.

"It's Juan," he said with a beaming smile. "I got a liver."

I looked closer at his face—the yellow was gone from his eyes and skin, and his gaunt features had filled out. I had only ever known Juan on death's doorstep, and that was how he became defined and immutably fixed in my mind. This man looked healthy with no hint of the protuberant fluid-filled belly I had examined each day. He stood before me completely on his own strength, the first time I ever saw him stand without the assistance of two people. He was transformed, his body restored by the magical cure of a working liver.

Juan reached out to shake my hand, and all I could do was hug him. It was one of the few times in my medical career that joy and relief made my eyes fill with tears. Behind him, his wife, Anna, gave me a tender, knowing smile.

Shared experience is a necessary ingredient in empathy, but medical knowledge adds something more to it. Though nothing objective in that hospital room could have indicated it, Juan, Anna, and I contained inside our heads a shared awareness and a memory. Though I had participated in only a snippet of the most difficult time of their lives, I knew of its morbid depths, and I had the faintest inkling of what they had been through to get where they were when we met again.

It is easy for patients to get lost in a hospital's shuffle. Doctors and nurses come and go, busy residents rush to and fro, and the constant cycling of sign-out makes physicians and patients alike as faceless as food separated from the living body that once fostered it. As a result, deep doctor-patient relationships in the hospital are rare. It is sometimes difficult to recognize the humanity behind a patient's immediate and pressing medical problem, or to see the life story behind their hospital course, and this makes empathy not always come so easily through the rush of the healthcare industry. I have forgotten about more of my patients than I can count, but Juan was one of the few who stayed with me.

When I see new patients with whom I have no personal experience, I can still know something of their story based on my experience with previous patients. People with failing livers carry evidence of the organ's dysfunction on their faces, and because I recognize those subtle signs, I can know what is going on unseen inside the body. Because I have experience shepherding patients like Juan and loved ones like Anna through the consequences of this unique organ's downfall, I know something about what a person has been through, and what they have in store.

When I meet patients who have received a liver transplant in the past, though they might be well and appear healthy at the time I meet them, I am aware of what they went through, the illness and pain their bodies experienced while hovering near death before being lucky enough to be transformed by a transplant. Though I have never experienced liver disease in my own body, I have borne witness to many who have been down the same pathological road. Having medical knowledge and experience as a physician means seeing deeper into other people's lives and grasping more than meets the eye, just as knowing where our food comes from means understanding the experiences of other kinds of creatures. The foundation of empathy, like taste, is not in the senses or the face's wiring, but in the awareness of what lies hidden behind life's superficial facade. Empathy for humans and animals alike merely means knowing what it is like to be alive.

With a patient like Juan, whose liver failed through no fault of his own, empathy comes naturally. But I have also had many patients who drank their own livers into cirrhosis, and they need compassion too. I find empathy to be easiest in pediatrics—whatever the medical problem, it is virtually never the patient's fault, while with adult patients, the opposite is usually true. Sometimes I have to try really hard to empathize—I've cared for murderers and child abusers, and I've learned to simultaneously feel both disgust and empathy for a patient. The nature of a physician's job makes victims out of everyone, no matter what they've done in the past. Empathy is not always easy, but it always matters.

6

PINEAL GLAND

On the morning of my very first day working in a hospital as a third-year medical student, I woke up at the ungodly hour of 5:45 A.M. The dark sky was barely tinged by sunlight on its eastern horizon as I showered and dressed, and my mind still felt fuzzy as I hurried through eerily deserted city streets that felt postapocalyptic in the morning twilight. By the time I reached the hospital, sunrise's half-light was finally over and the sky had lightened from the darkest blue to a lighter aquamarine, as if the day, and I with it, were finally coming up for air from the ocean's depths. I was about to start my very first rotation in internal medicine, and the sky's familiar daytime color felt strangely comforting.

Sign-out—when overnight residents hand over care of patients to the morning crew of residents and medical students, was at 6:30 A.M. sharp. I was not a morning person by any stretch of the imagination, but being late was a grievous crime—exhausted residents finishing their night shift would need sleep much more urgently than me, no matter how early I rose or how sorry I felt for myself.

My circadian rhythm training had begun.

I was assigned one patient—an old man hospitalized with pneumonia. After two years in lecture halls, this was my first attempt at playing a real doctor. I went to his room to take a history and do a physical exam so that I could present the case to an attending physician—the attending, for short—overseeing my work. I entered the patient's room as quietly as I could and found him sleeping in bed with oxygen tubing in his nose. In my hand was a clipboard with a ridiculously long list of questions I had printed out the day before, and I needed to get through all of them before meeting with my attending.

But I hesitated to wake the man. He was sleeping peacefully and snoring loudly as the soft morning light coming in through the window bathed half his body. I knew snoring could mean obstructive sleep apnea, but the monitor beside his bed showed a good oxygen level despite both the pneumonia and his snoring.

I tapped his foot several times, but the snoring continued unabated. I considered retreating and going back to ask my supervising resident how to proceed, but I knew he would just tell me to try harder. The residents had their own long list of patients to see before checking in with the attending, so I resolved to get through this one patient encounter on my own.

"Sir?" I said meekly. Nothing.

I shook his foot. Still nothing.

"*Sir!*" I yelled a little too loudly.

He awoke with a start. I introduced myself and immediately began peppering him with questions, a rapid-fire interrogation before the sleepy look had fully left his eyes. Once I finished with the barrage, I stripped off his blanket and examined him. I listened to his heart and lungs and pushed on his belly as I had been trained to do. It was my very first time in a doctor-patient relationship, and my first act was to wake an ill man from what appeared to be a deep and peaceful slumber.

After leaving him ragged and bleary eyed, I hurried off to rounds, and staggered through the rest of that morning in an exhausted mental haze. The early-morning punctuality fit me as poorly as my new work uniform—a collared shirt and tie tightened against my carotid arteries like a threat, and stiff black shoes assaulting my feet during hours of standing on rounds. Coffee was no longer a morning luxury but a pharmacologic necessity—in order to get through rounds, I depended on intermittent sips from a warm cup of my biochemical crutch.

My surgery rotation was even worse. Each evening before going to sleep, I set my alarm for the positively satanic hour of 4:30 A.M. Instead of the formal shirt-and-tie of internal medicine, I wore a pair of light-blue surgical scrubs, an outfit appropriately pajama-like for my half-asleep 5:00 A.M. train ride to work. Each morning I arrived at the hospital train station wishing the trip had been just a bit longer to give me more time to wake up. As I walked toward the hospital and another marathon of surgery rounds, the sky overhead was completely black with a few shimmering stars and not even a hint of sunrise.

Along with a pack of fellow medical students, I followed behind the attending surgeon and surgery residents as we roved the hospital's upper floors, waking one post-operative patient after another from sleep. Our primary question was whether the patients had begun to pass gas since the surgery, a sign their bowels were awakening after the expected postsurgical slumber. The patients barely had time to wake up before facing our questions and probing stethoscopes. Often, neither had their bowels—and for that matter, neither had I.

Surgery rounds had one consolation. On many of those mornings, I peered through my mental fog and past the repetitious fart talk and was dazzled by the sunrise. My tired eyes had trouble focusing on nearby objects and instinctively relaxed into the distance, where, through the patients' hospital room windows, they were rewarded by flaming blazes of color, bright crimson puddled on the horizon swirled with orange highlights. On clear days, each east-facing room in the hospital offered a breathtaking show over the landscape—a blessed distraction from the heavy ritual of surgery rounds. And so it was from the unlikely confines of an urban surgical ward in Camden, New Jersey, that I first learned to appreciate the morning spectacle offered by the sun, a fiery pill coughed up by the eastern horizon each morning and swallowed again each evening by the west.

A career in medicine clearly meant a tight morning schedule, and I wondered whether my body was biologically compatible. Through my teenage years and into college, I'd always been a night person: in the absence of any scheduling commitments, I'd often wake up around noon. In the parlance of researchers who study the circadian rhythm—the human body's cycle of sleeping and waking—I was an "owl," thriving at night like the famed avian hunter hooting unseen in the dark. At the other extreme are "larks," people who rise early and thrive in the morning, like the small bird whose operatic song often accompanies sunrise's first blush. People with a strong inclination toward one or the other theoretically gravitate to jobs and careers that fit their own intrinsic circadian biology, and I worried about my own career choice.

Adjusting my sleep schedule to the medical field would depend on my pineal gland, a pea-sized nub of flesh deep down in the brain's very center. The pineal gland is the body's organ of sleep schedules, and a lesser-known endocrinological cousin and nearby neighbor of the pituitary. The hormone melatonin drips from the pineal gland into spinal fluid a few hours before bedtime, and it prepares the body for sleep. The process happens much earlier in the evening for larks than for owls, allowing them to go to sleep early and wake up early, feeling refreshed—a feeling I could not appreciate.

As an owl trying to become a lark, changing my brain would require regular exposure to sunlight. When sunlight hits the human body, virtually all of it stops dead at the skin. Only through the eyes are the sun's rays able to sneak past our otherwise opaque covering and penetrate farther inside the body, and this is how its signal reaches the pineal gland. Light deactivates the pineal gland's melatonin secretion, signaling to the body that the time to sleep is over. And research has shown that, even more than a steady, bright source of light, gradually increasing illumination as the eyes experience during dawn is the strongest signal entraining the body's circadian rhythm. Though I did not know it at the time, watching sunrises on surgery rounds was helping to retune my pineal gland.

The training was working for me. As my time in medical school neared its own sunset, I no longer needed to be militant about bedtime in order not to feel completely brain-dead the following morning. Instead, getting to bed early became effortless, as if the medical profession's ruthless time frame had finally crept into my pineal gland's biology. Quiet empty streets around dawn, once eerie, felt completely tranquil with their own serene beauty. And on my days off, I would often rise with the sun despite no sign-out beckoning.

I had become a lark, and it opened my eyes in many ways. Doctors, it turns out, run on the same morning schedule as the natural world, and waking up in time for sign-out translated well into getting an early start on my days off for climbing mountains, fishing, and gathering wild edible mushrooms. The early hours are when everything in the forest is freshest, including mushrooms, and the best time of day to spot wildlife. As the sun rises, deer make their way back to bedding after a night of feeding and prancing in cornfields. They walk along accustomed trails on their own morning rounds, nipping at twigs and green buds, but at least they take care not to wake anyone else up while doing it.

I was proud of my newfound ability to wake with the sun, but my conversion from owl to lark, I later learned, is nothing special. According to Dr. David Dinges, a researcher of sleep biology at the University of Pennsylvania, many adolescents go through an owl phase that is probably more due to social cues than inherent biology—this would explain my love of sleeping late in adolescence. Dinges told me that most people are somewhere in the middle between the circadian spectrum's two extremes, and it was likely I'd always been a lark.

Much of what we know about the circadian rhythm, Dinges explained, is from studies that measure body temperature. Melatonin secretion causes the body to give off heat, which is why people sometimes feel a chill in the evening a few hours before bed. In the morning, as sunlight turns off the pineal gland, body temperature increases again, like the earth rewarmed by a newly risen sun driving away the nighttime cold. And sunlight's potency as a regular morning dose of strong medicine can help people shift the pineal gland's rhythm, as perhaps it did for me.

The daily cycle of sleeping and waking is the rhythm of most life on earth, a cycle we share with animals, microorganisms, and some plants. As the planet spins, its surface dwellers are flung into and out of sunshine repeatedly, and levels of hormones like melatonin in the bloodstream rise and diminish like birdcalls throughout the day and night. The pineal gland helps keep the musical meter of human anatomy and physiology synchronized with the outside world. Just as our feet connect us to the earth and our lungs root us in the atmosphere, our pineal glands conjoin every cell in our bodies to the sun.

Once I started residency, my list of patients ballooned from two or three at a time to ten, and sometimes more, and my morning rush intensified. Rounds with my attending, in which I would review each patient and their plan for the day, were scheduled for 9:00 A.M. Before that hour, I needed to visit each patient in their room, chat about how they were feeling, ask whether their symptoms had improved, discuss any side effects they might be experiencing from the medications I'd prescribed, and do a physical exam. Once I finished with all of that, I needed to review new laboratory tests and imaging studies, discuss my patients' pressing issues with subspecialty consultants, and look up information about their diseases and medications to fill the holes in my knowledge before meeting with my attending. Attendings often quizzed, or "pimped," residents in front of others during rounds, and because I never knew which obscure detail the attending would decide to harp on, I had to be knowledgeable about nearly everything.

My own "pre-rounds" began bright and early at six thirty in the morning, and on most days, I woke each of my patients from sleep. When I entered a patient's room to find them already awake, I was thrilled—not because it meant I could avoid waking them up, but because it meant they could begin answering my questions immediately without needing a minute to overcome their initial grogginess. After a quick series of questions and a rushed exam, my next task was to extract myself from the room as quickly as possible. When the patient had a complicated question or wanted to discuss some chronic issue unrelated to their current hospitalization, I tried gingerly to cut them off, my eye always on the clock.

Each day I struggled to get my work done in time for rounds at nine o'clock, and when I failed to accomplish everything on my to-do list in the morning, the rest of my day was guaranteed to be a harried game of catch-up. Any complication threw off my whole schedule. If a patient's condition was not improving, it meant I needed to spend time thinking more deeply about their case, considering other causes for their symptoms, and deciding on additional tests to run, medications to try, and consultants to call. All of it took time that I did not have.

When I finished pre-rounding, I rushed to the conference room, where I would discuss my patients' cases with the attending. Several other residents, each with his or her own list of patients to discuss, were usually already there, looking calm and ready for rounds. I wondered if they were just more efficient than me, or perhaps just more ruthlessly willing to wake patients up even earlier than I did.

Though sleep makes up only about a third of each of our days, it is the foundation of a healthy circadian rhythm. Reams of research demonstrate its importance for the entire body. Health outcomes of various kinds are worse for patients who are sleep deprived, including cardiovascular disease, obesity, and diabetes. Yet sleep was the one essential bodily process that I, and the hospital system in which I worked, most readily ignored in patients. A topic that seemed much more pressing to us residents was our own sleep needs. We read articles showing that decision-making becomes impaired and medical errors increase when we failed to get enough rest, and that this in turn impacted our patients. Still, while I had adjusted and

usually got enough sleep, there was little hope of my patients getting sufficient rest while in the hospital.

During a review of the medical literature, we discussed research that
showed an epidemic of disturbed sleep among hospitalized patients. Some
studies pointed to regular vital sign checks at night as being among the
biggest disruptors of patient sleep, and I recognized myself as part of the
problem. I often mechanically ordered these scheduled checks with the in-
tent of discovering a fever or alterations in heart rate and blood pressure as
an early sign of a patient's deteriorating condition.

Other studies faulted medication administration at times convenient
for doctors and nurses but inconvenient for patients—another detail of the
hospital's daily schedule about which I had never thought twice. When I
ordered morning medications for my patients, the computer algorithm set
a default administration time of 8:00 A.M., and I never considered changing it. After all, I myself was quite awake, caffeinated, and most of the way
through pre-rounds by that hour.

Some research pointed to artificial lighting in the hospital as being
problematic. The fluorescent lights shining in hallways throughout the
night gave false signals to patients' pineal glands and confused the physiologic onset of sleep in their bodies. Natural light promotes pineal health,
but hospital patients are exposed to very little of it, especially those farther
from the window in shared hospital rooms. For many patients, hospitals are
on par with casinos, which supposedly shield people from the time of day in
order to keep them gambling longer.

The most devastating consequence of sleep loss that I witnessed was
hospital-acquired delirium, which I encountered frequently in the elderly
and frail, especially those with underlying dementia. When my patients
became confused, disoriented, or generally mentally impaired, it often
stemmed at least partly from disturbances to their circadian rhythm. Delirium in the hospitalized has many causes, including disease and medications
themselves, but getting poor sleep in the hospital contributes to the problem. I was trained to focus more on treating the condition with intravenous antipsychotics, which do little other than calming crazed patients.

I recognized the problem and knew I was a part of it, but I needed to lower my head and rush through each grueling day in order to get my work done.

It didn't help with the epidemic of sleep loss that the hospital's cacophony of alarms squawked day and night. If the hospital were a natural outdoor setting, its birdsong would be the shrieking warble of heart monitors emitting false alarms and the forlorn, high-pitched call of a blocked IV pump ringing out from a distant patient room. Attendings and senior residents occasionally paid lip service to the sleep issue, but there never seemed to be sufficient time to adjust my practice or change my pre-rounding style to improve the situation for my hospitalized patients. My rushed resident brain was already masterful at tuning out the hospital's shrill soundscape, and I was quickly getting to the same jaded place of tuning out the sleep needs of my patients.

It was late in my first year of residency that I met Ted, a man in his mid-thirties who was admitted to my medical service. A few months earlier, a gnawing pain in his abdomen and some weight loss had led him to visit his family physician. A cascade of abnormal blood work and imaging tests followed, which led to a diagnosis of advanced stomach cancer. When I cared for him, he was already well into his treatment course of surgery, radiation, and chemotherapy. He'd had a fainting episode on the day he was admitted to the hospital, and his oncologist decided that a brief hospitalization for intravenous hydration and supplementation of his depleted electrolytes might help him feel better.

When I met him in his hospital room, he appeared gaunt, his eyes sunken and his blond hair thinned and falling out in clumps from the toxic treatments. His voice was so weak that I had to lean forward to hear the answers to my questions. I also met his young wife, Samantha, who was crying and distraught while clearly trying to hold herself together in front of their two stone-faced children, eleven and nine years old. In the face of towering and unstoppable tragedy, Ted's wife was understandably falling apart outwardly and emotionally, while his body quietly fell apart on the inside.

Over the following few days, I thought about Ted while showering before work, and I cringed at the thought of waking him up on rounds. Interrupting his sleep meant bringing him back to a reality that, I imagined, he had hoped would turn out to be a bad dream. Though I had gotten used to waking my patients, with Ted it felt very different. Perhaps it was because he looked so ill, and his case was so obviously terminal, that it felt even more patently useless to disturb his slumber than with my other patients.

Since his diagnosis four months earlier, he had been in and out of the hospital and his medical record was a torrent of physician notes, tests, and bad news. Reading through it, I felt like merely one more doctor in a line of faceless healthcare providers watching over Ted's terminal decline. I was just another stethoscope with a barrage of questions way too early in the morning.

Still, in an unthinking rush to get through my list before 9:00 A.M., I tapped gently on his leg until he opened his eyes. Each morning, I brought him back to the reality of my chilly stethoscope against his wasted flesh, my light shining into his dry throat, and my hands digging into his sunken abdomen. I imagined that every time, once the hazy confusion of sleep cleared, the terrible reality of his diagnosis would dawn anew, a daily horrific re-diagnosis. I asked about his appetite (always nonexistent) and his pain (constant and unchanged) before rushing off to rustle my other patients from sleep.

Though his brief hospitalization under my care seemed a slight detour on the road to his quickly approaching death, I nevertheless went through the motions of caring for him. I ordered IV fluids, futilely encouraged him to force tasteless food against the void in his gut, and pointlessly chased after his daily blood test results, supplementing his stubbornly low levels of magnesium, calcium, and potassium despite knowing they would simply fall once again shortly after I discharged him home.

Each morning, I expected him to stop me and say something like: "We both know I'm dying and there's nothing you or any other doctor can do about it. Is it too much trouble for you to let me sleep late?" But he never said anything of the sort. And in our quick twice-daily conversations, neither of us ever acknowledged the elephantine prognosis in the room. I left those discussions to his oncologist, who was primarily responsible for treating his cancer and who I knew was having daily in-depth discussions with Ted and his wife. The oncologist had known them since the original diagnosis a few months ago, a long-term relationship from the perspective of hospital medicine.

A few months after discharging Ted, I looked up his medical record. The very last note in his chart was written by his oncologist and contained a few sentences about the night he died—at home. I was happy that he died in his own home and not in the hospital. I pictured him sleeping all he wanted in those last few days, free from healthcare providers like me prodding him awake.

After my experience with Ted, I made a new rule for myself: I would never again wake a sleeping patient suffering from a terminal disease unless it was truly an emergency. Instead, I would let them dream of health and longevity for as long as possible. I could not apply this rule to every single patient under my care, though it would have been most humane, but I was becoming more efficient at pre-rounding, allowing me some freedom and a modicum of scheduling self-determination. I resolved to exercise my limited power in this way while still catering to the expectations of my attendings.

Instead of waking every patient before rounds, I let some sleep and based my initial plan for the day on information gleaned from the overnight nurses caring for them. I also tried to compensate for other sleep interruptions: I more readily offered my patients earplugs to protect against the hospital's incessant din, and I hung DO NOT DISTURB signs on their hospital room doors. I stopped thoughtlessly writing orders for vital sign checks throughout the night, and started considering which patients were actually at higher risk of becoming suddenly more ill. For the large majority, the risk was minimal, and I rescheduled their nighttime checks for saner hours.

Though the achievements of a single doctor in a massive sleep-annihilating healthcare system are limited, once I was an independent hospitalist, I was even more liberated to let both terminal and nonterminal patients sleep. Other doctors might wake their patients too early, too often, and for no good reason—but for my patients, I did whatever I could to let them sleep in.

The pineal gland was the body part I spent the least amount of time thinking about as a resident. It rarely gets infected, inflamed, or injured without a much larger problem threatening the brain as a whole, so I almost never needed to focus on it. People suffering from diseases of nearly every other body part appeared before me for treatment, but not once for diseases of the pineal gland. I had learned that tumors occasionally grow in the pineal gland, but I had never cared for a patient with that specific and quite uncommon disease myself.

Yet by better understanding the detrimental effects of sleep loss among hospitalized patients, I learned about the pineal gland's more subtle impact on health. Disturbed sleep in the hospital significantly impacts recovery from illness, but even for healthy individuals, our sleep is hampered by busy lifestyles and the false sunlight of smartphones, TVs, and computer screens. As a result, the body's natural rhythms are disrupted. The broad personal and societal repercussions of sleep loss are only now being uncovered, and a deeper understanding of the pineal gland might help us manage the problem better.

The pineal gland remains enigmatic—it was the last of the body's endocrine glands to have its function deciphered, partly because of its inaccessible position deep in the brain. Much about its far-reaching physiologic consequences remains to be learned: some research shows melatonin impacts the immune system, helping our bodies to fight infection, while other studies show it may even have tumor-fighting properties. Furthermore, taking melatonin tablets as a sleep aid proves extremely effective for some of my patients while having virtually no impact on others. The body's circadian rhythm is so deeply ingrained, and manipulating it, for better or worse, is a delicate craft that we barely understand.

Within hospitals, a new day is dawning: doctors, nurses, and healthcare systems are paying closer attention to the importance of restful sleep. Some hospitals are even implementing changes to reduce noise and increase natural light—interventions that are free and have no side effects, a rarity in the practice of medicine. Sleep remains one of the human body's lasting mysteries. Its importance and restorative properties are well known, but exactly how it refreshes the mind and body is still elusive—and sleep's guardian, the pineal gland, is one of the very few body parts about which we still have nearly everything to learn.

7

BRAIN

High in the Himalayan mountains, a headache is never just a headache—it is a sign the human body does not belong. While working in a medical clinic along Nepal's most popular trekking route, day after day I treated people with headaches, nausea, and dizziness. In a sea-level ER, these symptoms might have alarmed me, suggesting some intracranial catastrophe. But at altitude, the diagnosis was much simpler—the mountains and their colossal heights were making my patients feel so terrible.

I had taken the job in large part because of my love of mountains and travel, but also to learn more about the curious effects of an extreme environment on the human body. My trip began at low altitude in Nepal's hot and dusty capital of Kathmandu, where I received training in how to diagnose and treat altitude sickness. Well-being at high altitude is a peculiar niche of human health, one about which I learned virtually nothing in medical school.

More so than any other organ, altitude most severely effects the brain, which seemed fitting: the brain lives at the highest altitude of any internal organ—at least when we stand upright. But the brain is also our loftiest body part in other ways: it is the executive organ, the one that oversees all our other parts from its vaunted perch inside the cranium. The brain sits at the body's helm, pulling levers and steering wheels, a commanding captain with the human body as its ship. Even neurons, the branching electrical cells that make up the brain, seem more complex and majestic than any other type of cell found in the body. If organs were organized into a caste system similar to the strict social hierarchy found in India, and also in Nepal, the brain would be its Brahmin.

Yet the functions of the brain are still barely understood by the most advanced neuroscience research—it remains a black box, even though we live inside it. We don't quite know how consciousness is generated within this organ, and we continue to search for the precise point where the brain ends and the mind begins. But one thing we do know is that the farther we get from sea level, the more our brains tend to swell.

I f Mount Everest were a building with its first floor at sea level, its summit would be on the 2,900th story. I was assigned to work on the 1,150th story in the Himalayan village of Manang, located along the storied Annapurna Circuit. If a person rode an elevator from the ground floor to spend the night at that height, besides experiencing painful ear popping along the way, they would probably wake up in the middle of the night with the worst hangover of their lives. A thumping headache, nausea, and dragging fatigue are classic symptoms of acute mountain sickness (AMS), the mildest form of altitude sickness. Some would become still more ill from the elevator ride, the swelling in their brain reaching a life-threatening threshold, called high-altitude cerebral edema (HACE), which is responsible for most deaths from altitude sickness. (A large majority of people who succumb in the mountains still die of more typical bodily insults, like falling, getting buried in an avalanche, or having a heart attack or stroke.)

Altitude can also cause fluid to fill the lungs, but its effects on the brain cause most of the symptoms that people experience when they travel to high mountains like the Himalayas. How altitude causes these uncomfortable, and potentially deadly, changes in the brain's physiology remains a mystery.

By some unknown mechanism, the thin air, with its lower air pressure and reduced levels of oxygen, causes blood vessels in the brain to leak fluid.

The lectures I received in Kathmandu confirmed altitude's reputation for being, like the brain itself, a mysterious and poorly understood frontier of medical understanding. But despite the puzzle, the harsh manifestations of altitude sickness can still be mostly avoided. The risk can be minimized by allowing time for the body's acclimatization—gaining no more than sixteen hundred feet of elevation per day is generally a safe strategy. At that cautious speed of ascent, walking to Manang took me a week.

I started the trek with a fellow doctor from New Zealand, a Nepali translator and cook, and my wife in the warm tropics, where damp rice fields terraced the hillsides, and lizards flitted over sun-warmed rocks. Innumerable waterfalls thundering out of the clouds were the runoff from tremendous mountains above, though they remained hidden by fog. We gained altitude gradually, day by day, as tropical trees slowly gave way to temperate forests, and potatoes replaced rice in the fields. Toward the end of the week, birch and fir trees appeared, the same hardy stragglers of the Far North. Meat from yaks, an animal particularly suited to the high Himalayas, turned up drying over wood-burning stoves. Just as snow-covered mountains began peeking out from behind clouds, the dominant religion of local people faded from Hinduism into Buddhism, whose threadbare prayer flags, faded and flapping in the cold wind, told us we were finally at high altitude.

The Himalayan Rescue Association (HRA) medical clinic in Manang was two miles above sea level, with glacier-studded peaks rising another two miles just outside the clinic's door, straining both the neck and the imagination. The sweeping, steep-walled valley in which the village was nestled had all the makings of a real-life Shangri-La, the mythical high-mountain paradise ringed by snowy summits and inhabited by peaceful Tibetan Buddhists.

But my own symptoms when I arrived told me that tales of utopic Shangri-La glossed over altitude's ravages on the human body, especially the brain. More than others in our group, I had a headache, nausea, and poor appetite despite my slow and careful ascent, and I made my very first diagnosis of AMS on myself. Slow ascension is not perfect prevention—nothing is.

The HRA's medical director told us to think of AMS and HACE on a spectrum, with the amount of brain swelling determining the severity. As a physician, I was intrigued and mildly concerned to imagine my own swollen brain. As a naturalist, I wanted to learn how to better avoid symptoms of altitude sickness so that I could further explore the environs around Manang. And as a patient, I just wanted my head to stop aching.

When it comes to the brain in particular, a little bit of swelling can cause a lot of trouble, unlike with other organs. The lungs, for instance, are designed to constantly expand and contract, and their bony protection—a rib cage made of parallel bones with muscle in between—accommodates this movement by expanding and contracting in concert. The brain's bony protection, the skull, is much more rigid and immovable, especially after the first few months of life, when the mobile plates of a newborn baby's skull fuse into a solid, unstretchable encasement with very little room to spare. Even small amounts of brain swelling, whether from trauma, infection, tumor, or traveling to extreme altitude, can quickly crowd the skull and raise pressure inside it. This often leads to a rapid death as the brain's blood supply is strangulated and basic brain functions, like the control of breathing, falter. It's actually almost impossible to bleed out inside the skull, because a person dies from the brain being squeezed long before a decent volume of blood can be lost. There just isn't enough room in there.

Elevated pressure within the skull is treatable—a neurosurgeon can cut large windows into the patient's skull to give the brain room to expand without smooshing itself. Such a heroic measure probably could save some people from severe HACE, but it is not typically possible in the remote Himalayas, where there is no sterile operating room available and no fully equipped neurosurgeon. I was just thankful my own symptoms were mild.

Some aspects of altitude sickness made more sense to me when I pictured how the brain's appearance changes over the course of a person's lifetime. At birth, an infant's brain is chubby, its convoluted and folded surface smooshed up against the skull's inside with hardly any room left. A CT scan of a baby's head looks like a crowded subway car at rush hour. But as we age, our brains shrink, a process that is sped up in alcoholics and those suffering from strokes. In CT scans of elderly people's heads, the brain appears less like a plump grape and more like a shrunken raisin, with an abundance of room between its folds and significantly more space between the brain's outer layer and the surrounding skull. As undesirable as having a shrunken brain sounds, there is a benefit at altitude—elderly brains have more wiggle room to swell, which is why older people often suffer less altitude sickness than most young, fit trekkers.

My own AMS improved over the course of two days as my body adjusted to the new altitude. According to the current understanding of acclimatization, it was my lungs and kidneys that cooperated over that period to make the altitude bearable for my brain. Besides accepting the fact that the majority of my patients' brains were swollen in Manang, as was my own, I also had to alter my expectations of health in other ways: people's immune systems seemed to act fluky at altitude. Minor scrapes became infected more easily, and bacteria laughed at the antibiotic creams that usually kill them off quickly. The altitude also played games with these creams—they would unexpectedly squirt from their tubes when first opened. Produced and stuffed into their containers at lower elevation and then brought to higher altitude still unopened, those tubes were like the aching heads of tourists with altitude sickness, their internal contents bulging inside the skull's sea-level packaging.

Breathing also became a constant challenge, and my body never got used to the noticeably thinner air. Low blood oxygen levels, or hypoxia, which would normally have me rushing to give my patients supplemental oxygen, became routine and mundane. And my own was no better. At sea level, I identify a patient's breathlessness by seeing them pause to breathe in the middle of a sentence, but in Manang I, too, was winded in simple conversation. I easily lost my breath climbing eleven stairs to my bedroom on the clinic's second floor, and I panted even from the slight exertion of toweling off after a shower. What is pathological at sea level becomes normal high in the mountains.

In the yard just outside the clinic, tall marijuana plants grew in profusion, as they did all around town and virtually everywhere along the initial trek to Manang. As a naturalist, I was excited to see one of the most celebrated plants in the world in its native Asia, and it seemed to grow quite well at various altitudes. Marijuana has made its way from Asia to everywhere in the world that humans live for the simple reason that human beings, no matter their altitude, longitude, or latitude, enjoy chemically manipulating their brains to purposely cause malfunction, sometimes with very pleasurable results. Altitude seems to twist the brain from the outside in, but intoxicants like cannabis's active ingredient filter into the skull through the bloodstream and twist it from the inside out.

Intoxicants also help illustrate how different parts of the brain are dedicated to specific functions. Among its other effects, marijuana causes people to eat more—the "munchies"—because it affects a part of the brain called the hypothalamus, which plays a role in regulating appetite. Alcohol, another popular mind-altering substance, causes people to stumble because it disturbs the cerebellum, the part of the brain responsible for balance. The effects of any drug, as well as the effects of high altitude, can be traced to its impact on specific brain regions because the brain divides its labor geographically, with each part playing a different role in the organ's grand scheme.

There are many different ways to carve up the brain and understand its general organization. One such framework is according to altitude—the higher one goes within the skull, the more complex the brain's function and the closer one seems to get to the mind. Though the brain is at the top of the body's organ hierarchy, it contains within it an internal pecking order of its own.

The brain's lowest level begins at about the height of our nostrils. At this spot, all the nerves coming from arms and legs have coalesced into the spinal cord like a plant's branching root system forming its taproot—where that taproot first pokes above the soil and becomes a stem is equivalent to where the spinal cord first pokes into the base of the skull and becomes the brain stem, the bottommost part of the brain. This lowest level oversees basic bodily functions, including the foundational rhythms of heartbeat and breathing, tasks that are far from the loftier functions we usually associate with consciousness.

At altitude, the brain stem goes shaky: the normally steady, regular rhythm of respiration becomes erratic, with periods of rapid breathing alternating with pauses. This phenomenon was why people in Manang often came banging on the clinic door in the middle of the night in a panic, having woken up gasping for air. Like many of altitude's most unpleasant symptoms, it often struck at night—while sleeping, a prolonged lapse in breathing caused oxygen levels to sink even further, and people were startled awake by the feeling that they were suffocating. Most of these patients also had AMS symptoms, and I treated them for it, but they needed reassurance and an explanation for the distress they felt—the same sensations that many people with sleep apnea experience even at sea level.

Moving on from the brain stem and traveling to a higher altitude within the skull, one finds the brain's emotion centers. Structures like the amygdala and hypothalamus play a role in feelings of fear, apprehension, and anxiety, and it is at this level that the brain shows its first inkling of subjective experience. Altitude also impacts these brain regions, since lower oxygen levels are known to make people more irritable. Emotions have a component that occurs in the body, such as a racing heart and elevated blood pressure accompanying fear or anger, but their other side is a conscious awareness of feelings. Emotions straddle the body and the mind, just as they straddle the lowest and highest levels of the brain.

Traveling past the emotion centers to the skull's most extreme internal altitude, one comes to the brain's summit, the cortex. Only a few millimeters thick, the cortex is a veneer that covers the cerebrum's entire convoluted surface, and it is where the brain's primary processing occurs—its folded surface is meant to increase computing power. At altitude, malfunction of the cortex shows itself in people developing problems with attention, learning, memory, and decision-making. Stories from climbers of Nepal's highest mountains show the depth of cognitive dysfunction and disordered thought caused by such extremes, often with fatal consequences. The cortex is the highest-altitude layer within our highest-altitude organ—the brains behind the brain's operation. If consciousness is anywhere in the brain, it seems likely that it is here.

For the same brain-twisting reasons that people ingest certain drugs recreationally, those same chemicals can be very useful to physicians like me, and they help illustrate the brain's levels. In particular, the ability to temporarily turn off a patient's awareness—to sedate them into unconsciousness—is invaluable when performing painful procedures like straightening a broken bone, replacing a dislocated joint, or inserting a breathing tube into someone's throat. It is this same ability to turn off the mind that distinguishes modern surgery from torture. Physicians have many drugs to choose from in sedating patients, but the one I use most is ketamine because it targets only the higher functions of the mind. Most other drugs used to sedate patients shut down the entire brain from cortex to brain stem, which means they impact awareness but also depress breathing. If given in large enough doses, those drugs can cause respirations to cease altogether—a potentially fatal side effect.

When I give my patients ketamine, it turns off consciousness, but their lungs keep breathing and their blood pressure is maintained. This makes ketamine safer than many other sedatives, and means it is difficult to overdose when using it recreationally, but it also helps draw a clear line between the brain's lowest levels and its highest, its most basic functions and its loftiest abilities. Ketamine is in a class of medications known as dissociatives because they tend to dissociate consciousness from the body—in other words, they work at the precise location where the brain ends and the mind begins, usefully severing connections between the two, but only temporarily. While most intoxicants offer people an escape from the everyday life of their bodies and brains, ketamine does this literally. Drugs like ketamine help us understand the workings of the brain and each of its levels—including the higher mind—all of which go haywire at altitude.

Working in Manang's clinic, I saw a half dozen patients almost every day complaining of headaches, nausea, loss of appetite, and insomnia. In each case, I asked them how quickly they had traveled to Manang—if they had ascended rapidly, it would point to altitude sickness as the cause of their symptoms and obviate the need to look for other diagnoses. Altitude became another vital sign, along with the rhythms of heart and lungs, and as surely as I checked each patient's pulse, I asked them for a detailed timeline of their movement through the mountains. I diagnosed innumerable cases of AMS, both in foreign and Nepali trekkers, as well as in Nepali guides and porters, and I recommended the same regimen that had helped me: rest, medication, and simply allowing time for the body to acclimatize over a few days. It was the first instance in my medical career that I had suffered through exactly the same ailment as most of my patients, and I could empathize with them about how crappy they felt.

For many patients, AMS symptoms appeared after traveling to Manang along a new road that had been completed two years earlier, the first to pierce this remote Himalayan region. Jeeps plying the road's precipitous, fog-shrouded cliffs gain altitude more slowly than an elevator, but they still allow people to reach Manang in less than a day, easily outstripping the body's adaptations. Even locals living in Manang's stone houses developed altitude sickness when returning home by vehicle after brief visits to lower elevations. Living at high altitude over years causes long-term adaptations in the human body, primarily the production of more red blood cells to help deliver scant oxygen to cells, but the process of acclimatization that soothes altitude sickness wears off quickly—it takes only a few weeks at low altitude, no matter how long someone has lived high in the mountains. The new road brings economic benefit—before it was built, everything in Manang had to be carried to town on the back of a human or a yak—but it also makes it much easier and cheaper for people to gain altitude at unsafe speeds.

For each patient with altitude sickness, a crucial part of my physical exam was to test their balance. I had them stand at one end of the clinic's dark and cramped patient room, and asked them to walk in a straight line, touching heel to toe with each step, as the police might test a driver for alcohol intoxication. This crucial diagnostic test told me about the severity of a patient's brain swelling: when it had reached a critical threshold, pressure within the skull acted just like alcohol, impairing the brain's basic coordination abilities. Often the inability to maintain balance is the first sign that a person's suffering was due to not just AMS but the more dangerous condition of HACE.

The large majority of my patients passed the balance test, but the few who failed had become ill after continuing their trek beyond Manang. Most trekkers on the Annapurna Circuit were headed farther up the windswept valley and another mile into the sky to cross over a mountain pass called Thorong La, the highest point of the trek. It lay two days of walking from Manang, and the increasing altitude in that direction caused many people to develop new and worsening symptoms. Those who became more ill as they approached the pass turned around and came back to Manang to seek our help. The steady stream of woozy and wobbly patients coming back to the clinic from that direction spoke of much higher altitudes ahead, and it was these patients who were most likely to fail the balance test.

One night, a young Dutch woman appeared at the clinic door, and her boyfriend, who accompanied her, said she was talking crazy and acting drunk. She reported feeling as if she were on the deck of a rocking ship. A local villager who saw that she was unable to walk provided a horse for her to ride most of the way back to Manang. As I gathered the details of her symptoms, she held her head in her hands and complained of a severe headache. In normal medical circumstances, a drunken patient is simply that—a drunken patient. But at altitude, her symptoms were much more concerning, and seemingly more than just AMS.

She was already considerably improved by the time she got to us. Descending a few thousand feet to reach the clinic had helped her—the unsteady feeling had resolved and her heel-to-toe walking was fine when I tested it in the clinic. She still had severe AMS symptoms, but the signs of more significant brain swelling were gone. For those with altitude sickness, descent is the ultimate therapy, and one of medicine's few perfect cures.

One of the patients who visited the clinic was a seventy-eight-year-old woman called Ani, a name given to female teachers of Tibetan Buddhism. Ani came in every few weeks for a blood pressure measurement and blood sugar check, always dressed in a baggy saffron robe and a matching wool cap that covered her shaven head. She was one of several lamas living in the caves festooning the high valley walls around Manang, and her cave was an hour-and-a-half walk straight up from the clinic. She had lived there alone for the last thirty-eight years.

On one of my days off, I made the grueling walk in the hot sun to visit her cave. I found Ani tidying up her rocky abode, which she had managed to make feel homey—a few rugs covered bare stone ledges, and wooden ladders helped her get from one part of the cave to another. It was cozy, with stunning views up and down the valley. Her simple and austere lifestyle impressed me, along with her tolerance of the thin air—she moved agilely around the rocky surfaces and didn't seem to be panting, as I was.

Ani told me, via a translator, that she spends most of each day meditating. That was the main reason she lived so high in the Himalayas—the solitude and silence of her abode were essential for concentrating on her spiritual practice. Even a tiny village like Manang was too noisy for her, with its constant din of cows, yaks, children, and clanging bells. Situated high up on the valley wall, her cave was remote even by Manang's standards and offered a refuge from the bustle of society on whatever scale. The higher one goes in the mountains, the inherently fewer distractions, humanity being increasingly rarefied, like oxygen molecules in the air. I also imagined that slowing down and achieving the focused mindfulness of meditation was helped along by the extravagant respiratory cost of any exertion at this altitude.

When I asked Ani why she meditated, she answered, "Because I want to know myself." For most people, traveling to the mountains often means heading farther and farther from home, but, as Ani suggested, it can simultaneously mean traveling deeper and deeper into one's own brain, or perhaps closer and closer to the mind within it as one goes higher and higher.

I asked Ani if she believed the mind and brain were two different things or one unified entity. Two separate things, she thought. Like the brain, she expounded, the mind is inside the body, but it is not something you can point to with a finger. It doesn't have a shape or size or color.

I probed further, asking, "Where do you think the brain ends and the mind begins?"

She replied, "You'll have to meditate to figure that one out."

For a different perspective, I spoke to Dr. Benjamin Yudkoff, a psychiatrist and close friend of mine. In medical school, where we met, Yudkoff often got light-headed at the sight of blood, so the specialty of psychiatry—focusing on the mind rather than the body—was a perfect fit. But when I asked him about the human mind, he told me that he doesn't believe it exists at all.

At first glance, this seems as impossible as an internist not believing in internal organs, but Yudkoff made his case: though our experience of the world feels integrated into the united whole of awareness, it is actually an amalgam. Each area of the brain, and every one of its levels—from brain stem to cortex—contributes a portion of our awareness. The mind, he said, is actually a layered fusion combining the most basic reflexes from deeper brain areas overlaid with the loftier functions of emotion and cognition.

I asked Yudkoff how a psychiatrist can evaluate and treat disorders of the human mind without believing it exists, but this posed no paradox for him. The mind, though stitched together, is just a way of understanding the manner in which various parts of the brain work together. *Mental illness* is shorthand for aspects of the brain's function that cannot yet be better understood with blood tests, CT scans, or MRIs, tools regularly used by neurologists to diagnose other forms of brain disease. Even looking at a brain biopsy under the microscope would not help a psychiatrist. Instead, the brain's mental aspects are primarily evaluated through conversation, the primary diagnostic for psychiatrists like Yudkoff. Sometimes the simple act of talking with a person even allows psychiatrists to save them from the ultimate act of mental illness: suicide, in which the brain kills itself and takes the rest of the body with it.

The mind is a necessary concept for psychiatrists, but Yudkoff admitted that it is a quaint and outdated idea. We've known for over a century that the brain divides its labor into various parts, and therefore, consciousness is similarly pieced together like a quilt, no matter how fluid it feels. *Mind* is simply the word we use to describe what it feels like to have a human brain, our internal organ of cognition.

Yudkoff told me that one day in the distant future, when we know significantly more about the brain and mind, neurology and psychiatry will have more in common as medical specialties. In that omniscient future, we may no longer have to wonder how it is that our brains exude awareness just as any other gland in the body secretes its biochemical product.

Before I left the Himalayas for good, I decided to climb to the summit of Thorong La. I followed Manang's river west out of town and walked for two days before arriving at the very last establishment before the pass. No one lived there permanently—there were just a few rustic hotels perched on a steep rocky slope, catering exclusively to trekkers trying to get over the pass.

I began the next leg of my hike early in the morning, and I was panting from the start. Every breath I took contained less than half as much oxygen as at sea level, and I needed to muster every ounce of energy I had just to pick up my foot and place it down for each next step. I could only walk in slow motion. The mountain views, on the other hand, were staggeringly beautiful in perfectly inverse proportion to the punishing drudgery and air hunger of high-altitude hiking required to glimpse them.

At the start of my hike, I had passed a few clumps of grass, but as I gained altitude over the next few hours, the last vestiges of life disappeared. When I finally made it to the pass, I found an otherworldly landscape dominated by rock and ice, with barely any oxygen in the air. There were no signs of life besides a few other out-of-breath tourists. Standing six hundred stories above Manang, I felt as if I had reached the end of the world. My body, and especially my brain, felt utterly out of place in such an inhuman realm—ice formations hung off the mountains' every craggy shoulder, and those shoulders seemed to be perpetually shrugging at the breathlessness of trekkers like me. They even shrug at the human lives that were lost near where I stood on the summit of the pass, back in 2014.

Humans live roughly two-dimensional lives on the earth's surface, and for much of history, the upward direction has been unachievable, a realm of inaccessible mystery. Flying was technologically impossible, and the mountains offered the physiologic stress of altitude, a danger that probably contributed to people seeing them as a foreboding domain, but also one with magical potential. Even the soaring domes inside a church, through their height alone, give worshippers the feeling of being in the presence of a supernatural power. And those domes are eerily reminiscent of our craniums' vaults, underneath which sit our brains: the internal organ of spirituality.

I thought about Ani and why religious seekers often go to such austere places as the high mountains, besides the peace and quiet found there. Many religions from around the world hold mountains in special regard. In my own Jewish tradition, the prophet Moses climbed a holy mountain to find God and receive the Ten Commandments, and various other religious traditions feature a mountain in their creation tales. Mountains are inherently remote and isolated, and naturally more pristine than down in the valleys, where people tend to live in larger numbers. The air is cleaner, in addition to being thinner, and waterways are less spoiled, since the waste and effluent from human bodies and human societies alike tend to run downhill. Going to the high mountains necessarily means getting away from society's defilements and the dirty fray of daily life, just as going higher in the brain takes one further away from the basic, animalistic business of the body and closer to the angelic mind.

People have wondered about the nature of consciousness since time immemorial, and there are innumerable philosophies and theories that attempt to explain it. To me, it seems likely that our minds spring from interconnectedness. The brain is composed of neurons communicating with one another, and these connections between cells are the basic unit of brain function. Taking a step back and looking from a higher vantage point, we can see that interconnected neurons are bundled together to form each geographic area of the brain, each with its own specific function, and it is through the interplay among these various regions that the consciousness of each individual person is built—a conversation among parts from which emerges the whole. And from an even higher bird's-eye view, a conversation between two people—a verbal connection between two brains—can reveal the mind's inner working. Perhaps traveling even higher and looking down from the highest heights, one might attain Ani's perspective.

For the same reason, getting "high" on drugs sometimes offers its own useful perspective on life. From the celestial heights of the Himalayas, as well as from more figurative cerebral heights, our daily problems seem minuscule and irrelevant. Altitude is therefore the perfect place for contemplation and meditation. Perhaps the strain of altitude on the brain also offers a helpful crutch to people seeking to free the mind from its terrestrial abode in the brain. Maybe spiritual seekers in the mountains are just getting high—on hypoxia.

Some of altitude's most precious gifts are delivered only upon descending back to lower elevations. After two months of meager mountain air in Manang, I traveled back toward sea level in a Jeep along the new road, and as I descended, every breath felt fuller. For the first few days after my return, the richness of a full atmosphere felt luxurious—almost euphoric. I had forgotten the feeling of being simply unaware of my own breathing for days at a time. The aloof and removed perspective on life gained in the mountains remained with me for some time, before everyday life whittled away at it. And over the same time period, an everyday altitude undid my body's fleeting acclimatization to the high Himalayas.

Whatever our altitude, we look out on the world from our own personal mountains—the highest and innermost organ, the brain. Though it is not actually buried so deeply within us—it resides beneath only a thin veneer of scalp and skull with hardly any separating flesh at all—it provides the inscrutable vantage point from which we experience the world. Our brains are, therefore, the seats of our deepest selves. And, like the body as a whole, the brain is far more than the sum of its parts.

8

SKIN

For hide tanners, senseless death is an opportunity for craft. This was just one of the many lessons I took away from a wilderness survival course I signed up for in rural New Jersey the summer after graduating from university. Ever since I was introduced to wild edible plants in Central Park, I had been eager to learn more about living directly off the natural world—and wilderness survival seemed like the logical next step.

The instructor for the hide-tanning lesson on the second day of the course was named Gary, and he played the woodsman's part well—a large scruffy beard tumbled onto his red flannel shirt as he kneeled on the forest's damp leaf litter next to a road-killed doe. Earlier in the day, another instructor had driven by the lifeless carcass not too far from where we were staying. He had pulled over and loaded it into his pickup truck to bring back as a demonstration piece for our survival course. The instructors were casual about this backstory, and picking up roadkill seemed to them nothing out of the ordinary. I found the whole thing simultaneously dumbfounding and thrilling. It was the closest I had ever been to a dead animal before.

Using a small pocketknife, Gary cut through the doe's skin in a circle around its wrists, ankles, and neck, then up the inside of each limb. Another cut went straight down the belly, connecting all the other cuts. He set his knife aside and began forcefully pulling at the deer's skin with his own leathery hands. With repeated tugs, the skin and its amber fur coat fitfully released their grip on the red muscle and white connective tissue revealed beneath. The skin cleaved off the animal's body as easily as a banana's peel—or, as I had learned earlier in the course during a lesson on making hunting bows, as cleanly as a tree's bark in early summer. I had never seen an animal skinned before, and to my surprise, it was quick and bloodless.

Once the skin was freed from flesh, it looked like a large cloak. Gary poked holes along its outer perimeter and laced it into a large rectangular frame made of wood, where it was stretched in all directions like a trampoline, furry on one side. With a group of students following behind him, he carried the frame to a sunny spot and leaned it against a tree, where it would dry over the next two days.

Later that afternoon, during a break from lessons on how to make basic tools from stone and antler, I walked over to the drying hide. Moments before skinning, it had enclosed a complex, cervine geometry of trunk and limbs, but now the skin lay perfectly flat. It had already begun to resemble a swatch of fabric—the goal of tanning, and something humans have done with animal skins since time immemorial.

The thick amber fur felt coarse as I ran my fingers through it, and I noticed that each hair had subtle bands of white and black toward its tip—an excellent forest camouflage. I brushed away a few clinging pine needles. Walking around to the frame's other side, I hesitantly touched the skin's moist, hairless inner surface. Unlike finished leather, skin still fresh and hydrated from life felt rubbery, and it elastically sprang back when I pushed and released. A stray blood vessel snaked across it as flies explored the smooth sheet, looking for a promising patch on which to lay eggs on its rapidly drying surface.

Only just removed from the flesh it had once covered in life, that skin was the corporeal perimeter within which a deer had lived its entire life. I looked more closely at a small scar that Gary had pointed out, a whitish rough patch of skin that had lain over the animal's back. That scar told of an incident from the deer's life—perhaps an encounter with a sharp branch, barbed wire, or an aggressive buck in mating season.

Two days and several wild edible plant lectures later, the skin was dry, stiffened and rigid like cardboard. A new instructor, Lorie, picked up the hide-tanning lesson where Gary had left off. She wore a pair of overalls and a braid draped in front of each of her bony shoulders.

"We call this rawhide," she said as she flicked a finger against the hide's hairless side. A faint, muffled gong sound reverberated. She held up a sharp steel tool that we would use for the next step in the tanning process: scraping. Lorie scratched the scraper's blade against the hide's fur in long steady downward strokes, and shreds of hair and thin ribbons of skin began falling and piling rapidly at her feet.

"Understanding skin's layers is the key to making quality buckskin," said Lorie, "and the key to *not* tearing your hide to shreds."

She pointed out the skin's many layers as she removed them: just beyond the fur was a dark layer, then another one speckled like salt-and-pepper, and next a yellow layer. As each new layer was exposed, she pointed out differences in color and texture that were indiscernible to my eye, though I nodded along with the other students. Finally, Lorie scraped through to a gleaming-white layer of skin. We had reached the dermis.

"That's when you know you've scraped enough," she said, breathless from the exertion. "We want to see this lovely white color across the hide's entire surface. This is the key—scrape too little, and your hide will never come out soft. Scrape too much, and you'll ruin it with holes. It's all about layers."

She flipped the frame and scraped the hide's hairless side, removing a layer of thin membrane that had once hugged muscle and tendon.

The problem with turning skin into usable fabric, Lorie explained, is that once separated from the rest of an animal's body, skin has only two possible fates—either it dries into rawhide and is protected from decay (but it gains the hardened consistency of wood); or it stays wet and supple (but it rots). Neither makes a useful fabric for clothing. For a dead animal's skin to live a second life as apparel, it must be made both dry and supple at once.

The key to solving this riddle came when Lorie pulled a pink, glistening lump of jiggly flesh out of a white plastic bucket: the deer's brain. It was the size of a large orange, much smaller than the cantaloupe-sized human one I would dissect years later as a medical student. She held the brain tenderly in her cupped hands for everyone to see, careful not to let her braids brush across its wet convoluted surface. Brain, she told us, was the essential ingredient to magically transform a dank, smelly hide into luxurious buckskin—a process appropriately called brain tanning. The resulting soft material was the same that had clothed people throughout prehistory, as well as in colonial America.

Lorie returned the brain to its bucket and pounded at it mercilessly with one hand. She kneaded it and squeezed it through clenched fingers, occasionally adding a dash of water, until it transformed into a thick strawberry milkshake of mashed-up neurons. She smeared the pink muck onto the hide, which had been unstrung from the frame and lay on the ground, coating each side generously. Lorie then carefully folded up the hide and placed it back in the white bucket. Flies hovered in frenzied anticipation.

The following day came the transformation that astonished me. Lorie removed the wet hide from its bucket, wiped excess brain from its surface, and wrung it out with a firm twist. She then strung the hide back into the wooden frame, and the softening step in the tanning process began. Under her instruction, the other students and I took turns kneading the hide and massaging it with our hands, poking at it with sticks, and rubbing its surface with a rough piece of sandstone. She told us that we needed to keep the skin's fibers continually and aggressively in motion as they went from damp to completely dry. And we could not let up for too long, Lorie warned, or we'd be left with a piece of rigid and unwearable rawhide.

After several hours, the hide was completely dry, gleaming white in color, and as buttery soft as the finest suede. As Lorie untied the buckskin from the wooden frame, she listed some of its limitless applications—clothing, pouches, knife sheaths, bags, arrow quivers, hats, gloves, string, upholstery, and more. Skin's versatility impressed me, but I was mesmerized by the remarkable metamorphosis in which I had participated. Such repugnant initial ingredients had turned into a luxuriously soft and strong material with an impressive variety of everyday uses. I marveled that two organs from one smashed animal's body found on a roadside could combine to create a fine fabric so unthinkably distant from its odious anatomical origins. I was hooked.

I came away from the wilderness survival course with basic know-how in a variety of ancient skills, but tanning was my favorite. Skin as a medium for craft became my singular fascination, and I practiced brain tanning over the following months and years, right into the start of my medical education.

All human dissection necessarily begins with the skin too. My cadaver in anatomy lab was an old man lying facedown on a metal gurney, prone like a deer on the forest floor ready for skinning. His skin was cold to the touch, and the bloodless gray of thinly overcast skies. The medical school's mortician had pumped his body full of a smelly preservative solution that had kept his skin and the rest of his tissues moist and supple, but without rotting. He stayed in that preserved state for the several months it took to dissect him. Lorie had failed to mention a third fate for skin after life—preservation with industrial solvent.

Following instructions in the dissection guide, my fellow students and I started our first cut just below the nape of his neck, and the scalpel ran straight down the middle of his back. We made several horizontal cuts that crossed the initial vertical cut, and then began pulling loosened flaps of skin to both sides. The skin was greasy from the man's fat, slippery, and not nearly as stretchy as fresh skin.

Once we got the skin pulled out of the way, we ignored it for the remaining four months of anatomy lab. We studied the muscles, nerves, blood vessels, tendons, and bones hidden beneath the skin, but the organ that had caught my interest through tanning was treated as little more than an inert layer of wrapping paper to be torn aside and tossed away on a mission to reach the prize inside. In anatomy lab, it seemed as though skin were the body's superficial ornamentation, as irrelevant to the practice of medicine as clothing itself.

At the end of each day of dissection, we flipped the skin flaps back onto the muscles and organs beneath to prevent them from drying out—keeping the body moist is an important function of skin in the living too. Still, despite being covered in plastic, a few edges of the skin dried out over months, turning into recognizably strong and stiff rawhide. Bits of muscle, too, dried into stiff jerky, and I understood that our entire bodies have the same two fates after death that Lorie had laid out: drying out or rotting.

It was in histology class the following semester where I learned that skin is much more than a static outer lining. Skin is the body's largest organ, and it has a complex life of its own—it comprises an active and bustling layer atop the body like topsoil overlying the earth. Sweat glands burrow within it and help cool the body by secreting sweat to moisten skin. Neighboring sebaceous glands condition and lubricate skin by secreting sebum. Skin's embedded follicles sprout hairs, each equipped with a tiny muscle that stands the hair up on end in response to the frigid chill of winter, a creepy feeling, or mesmerizing vocal harmonies.

And skin is intelligent: as the only organ regularly exposed to the sun, it tans in response to sunlight, darkening with pigment to protect mutation-prone DNA from ionizing solar radiation. Skin thickens into callus in response to repeated rubbing insults, growing its own armor to protect against future friction. While the liver is an organ renowned for its regenerative capacity, as immortalized in the myth of Prometheus, I found liver had nothing on skin's ability to heal and renew itself. Skin wounds close on their own as cells invade from all sides to fill a traumatic defect.

I studied hundreds of pictures of skin's cross section magnified under the microscope. The top layer, called the epidermis, appeared as a thin outer veneer lying atop the much more substantial dermis, which made up nearly the entire thickness of the slice of skin. The outermost layer that we see and touch each day was only the tip of skin's iceberg.

But when magnified under a strong microscope, the skinny epidermis itself flowered into five different sub-layers like a stack of sliced cheese.

The outermost forms a hornlike and waterproof crust using the same material—keratin—that makes up our hair and fingernails. Each successive layer beneath it has a specialized task to carry out in maintaining skin's integrity and health: one layer fastens the cells of the epidermis together, while another produces the keratin. And the bottommost layer of epidermis—the one bordering the dermis—contains stem cells that divide as needed to replenish skin, fill wounds, and replace the dead cells that we shed throughout our lives, surrendering our skins as dust to our homes, cars, and workplaces. These epidermal sub-layers were the ones Lorie had pointed out years earlier, the signposts for scraping a hide just enough, but not too much.

Like skin, every part of the human body can be broken down into layers: the whites of our eyes contain four layers, artery walls have five, and the lining of intestines has six. Even the brain's thin cortex actually has six layers within it. Layering is a basic underlying architectural principle for the human body, and it provides resilience and allows for more specialization of cellular function. Every structure in the body, no matter how thin and simple it initially appears, is layered like an onion.

In my second year, I learned pathology—the study of disease—and quickly discovered how viscerally appalling disorders of the skin could be. Photos of blistering, ulcerating, and sloughing rashes were some of the most gruesome we saw that whole semester. One particularly abhorrent photo showed the back of a man's neck with a skin infection dripping with pus and maggots. According to the lecturing dermatologist, the man's infection had progressed over weeks before he came into her office to get it checked. The photo elicited a collective groan from the lecture hall.

One student asked, "How could someone let it get that *bad?*"

The dermatologist's answer: "Alcohol helps."

Internal organs hidden within the body were still unfamiliar to me when I began my medical training, so I had little context by which to feel revulsion at pictures of their diseases and infections, no matter how putrid. It took years of getting to know innards before I could be horrified by photos or CT scan images of their afflictions. Skin, on the other hand, was an everyday fact of my life, and seeing its most destructive pathologies provoked an innate disgust in me. Perhaps this is why skin diseases like leprosy or disfiguring burns have carried such profound social handicaps throughout history.

I began learning to decode human skin, each pathological image helping me to read the splotchy impressionistic alphabet of rash. When skin eruptions displayed tiny fluid-filled bubbles, or vesicles, I learned to think of chicken pox, Coxsackie infection, or herpes, while a rash of flat purplish spots could mean a bleeding disorder or deadly meningitis. Precisely where on a person's body a rash appeared had meaning too: when it colored a person's palms and soles, I was taught to suspect syphilis.

Even a rash's manner of spread gave diagnostic information. Asking patients about where it started and how it had expanded since that time was crucial: a measles rash typically begins on the face and spreads down the body like a window shade slowly being pulled down, while the rash of Rocky Mountain spotted fever begins on arms and legs and spreads inward as if beckoned by the trunk's gravitational pull. Sometimes feeling a rash with my hands told me its cause, such as the characteristic sandpaper feel of a scarlet fever rash or the raised purple dots indicative of blood vessel inflammation.

And, of course, another important part of my training came from the endless barrage of text messages from friends and family members containing poorly lit photos of what were usually simple insect bites. Over the years, I gained more experience diagnosing rashes, and I began to more quickly recognize patterns inscribed on skin's paper-like surface.

A strange rash on a child that I saw while working in Arctic Alaska after residency required me to rely on my ability to read skin's layers to make a crucial decision. An Iñupiat woman brought her one-year-old daughter into the ER because, after several days of fever, her skin had begun to peel. Once the child was undressed, I saw thin, translucent sheets of skin flaking off her chest, arms, buttocks, and much of her face. It looked as if she had been badly sunburned, but it was late autumn in the Arctic, and the sun barely rose above the horizon for a few hours at midday. When I rubbed my hand vigorously against areas of skin that had not yet begun to peel, small bubbles formed and quickly grew and coalesced into larger expanses of shedding skin.

Two possible diagnoses came to mind: the first was a deadly condition called Stevens-Johnson syndrome (SJS) where the skin's epidermis is completely sloughed, leaving only dermis on much of the body. Patients with SJS often require intensive care in specialized burn hospitals because of the life-threatening danger resulting when skin loses its outer layer. The other possibility, staphylococcal scalded skin syndrome (SSSS), was much less deadly, and resulted when a toxin disrupts the glue holding sub-layers of epidermis together. In SSSS, only antibiotics are needed to treat the underlying staph infection producing the toxin.

When I looked closely at the child's skin, beneath the sloughed layers I saw dull pink, dry-looking skin—these were most likely deeper layers of the epidermis, rather than the beefy red color of living, inflamed dermis. I therefore suspected SSSS, and my hunch was supported when I found that her rash did not affect the specialized skin of mucous membranes in her nose, mouth, eyes, and genitals, a finding expected in SJS. I held off on an immediate evacuation to a larger hospital in Anchorage, and the child improved dramatically over the following two days with antibiotics alone. As Lorie once said over a bucket of brains, "Understanding skin's layers is the key." Little did I know then that this piece of advice would serve me well beyond hide tanning.

Over the years since the wilderness survival course, my skill in tanning hides improved. I even became comfortable picking up roadkill to harvest its skin, and I carried a knife, gloves, and a large black garbage bag in my car's trunk at all times. One summer afternoon, while driving in upstate New York, I passed a deer lying on a two-lane road bordered on either side by forest. I pulled over and got out of my car, and once I determined that the animal was fresh and unspoiled, I dragged the body away from the roadside into the forest, where other cars wouldn't be able to see what I was about to do.

I pulled out a small knife and skinned the animal using the technique I had learned from Gary. I was getting faster at it, and I never tired of the act of pulling an animal's skin off—it was as satisfying as peeling dried Elmer's glue off my own skin as a child. I harvested the animal's brain as well, using a small saw, and placed it in a separate plastic shopping bag.

At home, the deer's skin dried in my hide frame over the next two days, and then I began scraping it with a scraper I had made from an old piece of discarded steel. I could finally recognize the layers of epidermis that Lorie had been trying to point out years earlier—the same ones I came to understand in medical school. As I scraped through each of them, I thought about the epidermis's appearance under the microscope: it looked like cells stacked on top of one another like a brick wall. I carefully removed each layer of bricks until only white dermis was left.

Within a few hours, I had the hide fully scraped, smeared with brains, and soaking in a bucket. The next day I spent several hours in the sun, softening the hide strung into the frame, my own sun-dappled skin growing sweaty as the deer's skin slowly softened under my manipulations. The skin was tough and easily resisted my constant tugging—I knew that its strength and elasticity wholly reflected the dermis, which under a microscope looks like a tangle of strong collagen fibers woven into a solid fabric.

That deer hide came out softer than any I had yet tanned. I cut a small rectangle off the lower back of the finished product, dyed it with henna I bought from the store, and sewed it into a small pouch that I filled with all the medications I might ever need while traveling. I called it my "travel pharmacy," and to this day, I rarely leave home without it.

For each and every one of my patients, skin has been the first organ that I've examined through visual observation, and often the main organ with which I've interacted. Whenever I push my hands into a patient's abdomen, feeling for enlarged or inflamed organs, my hands actually touch only skin. My stethoscope presses against skin when I listen through it to hear the heart, lungs, and bowels. And whenever I manipulate an injured shoulder, gliding it through a normal range of motion and checking for fractures, strains, and ligament tears, I only ever see and feel skin, never the shoulder joint itself. In the same way, we see only skin when recognizing a familiar face, though its shape depends more on the architecture of its underlying bone, cartilage, and fat. My patients only ever come wrapped in skin, and skin is the primary portal through which I assess their internal health.

Despite being a uniquely external organ, skin offers important clues to physicians about the well-being of a patient's hidden insides. Jaundice tells of liver disease, while brown and thickened skin over a patient's lower legs speaks of chronic heart failure, with especially severe cases turning human skin nearly into the consistency of tree bark. Even some cancers hidden within the body can manifest as a velvety black rash or a lacy pattern over both eyelids.

Skin's intimate connection to our insides is particularly useful in assessing the brain. I once evaluated a patient with a numb left leg, along with weakness and other symptoms, and when I lightly grazed my fingers against the skin of that leg, he felt nothing at all, and I suspected a stroke on the right side of his brain. The skin's sense of touch provides a convoluted surface of pure perception that envelops each of us, and a touch or texture anywhere should trigger a brain response, an awareness. But this patient's sense of touch was disrupted, and I knew that that specific patch of skin on his leg was wired to a corresponding patch of cerebral cortex on the opposite side of his brain. An MRI soon confirmed a stroke in precisely that location. The body's entire skin-covered surface is mirrored directly onto the brain, and by simply touching my patient's skin, I could delineate the brain's well-being and precisely locate any hidden insult, like strokes, brain tumors, infections, or hemorrhage. Skin's externality helped me to interpret internal matters.

Sometimes when I first walk into a new patient's room, one glance at them gives me an uneasy feeling about their clinical condition. I often hear nurses and doctors saying that a patient "just doesn't look good," meaning they are concerned that something serious is going on inside a patient without being able to quite put their finger on exactly what raised the suspicion. As my experience as a physician grew over the years, I realized that this feeling often comes from subtle visual clues of color from a patient's skin. Barely perceptible hints of gray, blue, and green often appear in the sickest patients, and I only began to see these colors once my eyes were trained to recognize them. Skin's palette of pathological hues often gives the first alarm bell for life-threatening illness.

Skin does a fairly good job at protecting the body from a treacherous outside world, but it often proves no match for knife blades and coffee table corners. A violation of skin's fundamental intactness, called a laceration, accounts for its most basic disorder, even simpler than rashes and off colors. Skin lacerations, each a break in the body's interface with the outside world, are extremely common, and closing such ruptures is a large part of my job working in the ER. Repairing lacerations, especially the deepest ones, is when my job practicing medicine most resembles crafting finished buckskin.

The worst lacerations I ever saw came into my ER late one night on the face of a young woman in her twenties. Blood-soaked gauze covered her forehead, eyes, and cheeks, and only her mouth was visible. She whimpered drunkenly. The paramedics wheeled her gurney into one of the patient bays, and as I began slowly unwinding the bandages wrapped around her head to survey the damage, a paramedic recounted her story.

She had been drinking along with a group of rowdy friends earlier that night. While standing on her home's porch steps, she got into an argument with her sister. Suddenly her sister pushed her, and she fell off the steps, plunging face-first onto the lawn below. In an unfortunate coincidence, a few days earlier her slow cooker had broken, and she had tossed it onto her lawn in that same spot. The slow cooker's cracked ceramic insert lay alongside her porch steps, and when she fell, her face crashed directly into the ceramic shards. I knew from my survival course that broken ceramic is as sharp as the fractured flint edges of prehistoric stone tools—the instructors had recommended practicing the art of flint knapping with broken, discarded toilets.

As I removed the final turns of bloodied gauze, I saw three deep lacerations crisscrossing her face. The wounds had mostly stopped bleeding by the time I examined them, as they often do—lacerated skin edges expertly stanch the flow of blood by narrowing blood vessels and clotting them off. A crescent-shaped cut split the middle of her forehead, while a shorter one traced the underside of her left eye, just missing the lower eyelid. But the deepest of the three injuries was a yawning gash to her right cheek. Once again: alcohol helps.

I explored the depths of each laceration as she cursed loudly. I pushed apart the edges and peered inside to determine how deep they went, ensuring that no underlying structures like tendons, nerves, salivary gland ducts, or large blood vessels were damaged. I also searched for dirt and ceramic shards that might need to be removed.

But most important, I was looking for layers. The small laceration under her left eye clearly cut through both epidermis and dermis, and into a layer of yellow fat beneath her skin. The crescent laceration on her forehead went deeper still, reaching beyond fat and into a glistening red layer of eyebrow-raising muscle. But peering into the deep chasm over her right cheek I could see even farther, past pink subcutaneous tissue and red muscle, all the way to a white layer that felt hard when I pushed against it with a sterile probe. It was the lining of the cheekbone itself.

Every laceration, every skin break, is a view into the body. For physicians of ancient Rome, wounded gladiators were their best opportunity to study the body's internal anatomy, since autopsies were prohibited. The deeper and more gruesome the wound, the more educational. For me, peering into wounds feels like walking past an urban construction site where digging has revealed the underground layers of a city normally hidden from view by its intact skin of asphalt. I often pause to enjoy such brief but captivating glimpses of a street's hidden pipes and wiring coursing through earthen flesh like blood vessels and tendons deep in a laceration. Knowing the body's layers is important for assessing lacerations, as is knowing a city's subterranean layout for urban planners digging down through a street to fix infrastructure.

When I read the body's layers inside my patients' wounds, I evaluate their extent, but I also plan my repair. Though small, shallow lacerations can close themselves, larger gashes like the ones this young woman had would need assistance from stitches. While examining her face, I pushed the edges of each laceration together to visualize how stitches might hold them closed. For the two deeper ones, pushing their edges together required more force than usual, and I knew it would take very tightly pulled sutures to hold them shut. Repairing such complicated lacerations under significant tension requires a tanner's and a histologist's understanding of skin.

After flushing the wounds with a large volume of sterile saline, a physician assistant and I placed the first set of sutures into the deepest layers of muscle, but the next row would hold the majority of the wound's tension. I knew the weak epidermis could never handle such strain—I would need to anchor the sutures in the dermis. I looked closer at the layers of her skin's cross section: beneath the suntanned, freckled epidermis I could see the dermis shining a pure white. The same layer of skin that gives buckskin its strength would provide resilience in repairing the most gaping wounds.

As I threaded the curved needle through her skin, I made sure that it pierced precisely through the dermis. My hands were practiced from sewing buckskin over the years, and I had developed a deeper intuition for skin's many layers. The main difference between suturing a tanned hide and closing up a living human's lacerations was that with people, I could not pull the needle through the other side of their skin. This is why the needles used by doctors are curved instead of straight.

Once we finished the dermal sutures, which would be buried inside the wound and dissolve over weeks as the dermis grew back together, the wound was nearly closed. We placed a final row of sutures into the epidermis itself—this layer offered no strength in keeping the wound closed, but was essential only for an optimal cosmetic result.

It took over an hour to finish, and the patient hardly made a sound—I was thankful she had come in already pre-anesthetized by alcohol to make the sewing easier for all involved. We pulled away the sterile drapes, stood back, and examined our handiwork. I imagined the horror she would experience the following morning upon first looking into the mirror. Her scars, though hopefully minimized by our careful repair, would forever tell a story about that night's encounter with a slow cooker, even if her brain could not quite remember.

Skin always carries the story of how a human or animal wore it in life—a story written in scar. Injuries from the past leave a mark, and as a tanner, I had learned to read scars to avoid weak spots in a hide, knowing my scraper blade could pop through them and rip a large hole. As a physician, reading human scars helped me in other ways—it told me of my patient's past run-ins with a surgeon's scalpel. On the skin overlying a patient's abdomen, scars gave me hints about possible causes of a person's abdominal pain. For example, a long diagonal scar across a patient's upper right abdomen told me their gallbladder had been removed long ago, and so gallstones were unlikely to be the cause of their discomfort. Any surgical scars at all meant that a person's abdomen was not virgin—it had been violated by a surgeon at some point, and the footprints they left would forever tell the tale of that operation.

But skin's story continues after death through the transformative process of tanning. My travel pharmacy and I have been through a lot together, traveling to four different continents over many years of my career as a physician. It has become as personal as my battered wallet. Near one of its lower corners, the buckskin pouch has a scar streaking across it like a comet's tail—a mark that might have been made during the deer's life, or by my scraper blade during the skin's second life as a fabric. When I uncinch the pouch and pull from it some medication—or earplugs, which I always carry to ensure better sleep while traveling—I feel the scar's rough surface against my own skin and I remember the road-killed deer it came from. A deer's tanned skin, like a cadaver's preserved body, provides a brief but useful postmortem pause on its way from life back to ashes and dust.

The pouch is weathered now, and my own skin, too, has accumulated more scars, wrinkles, and age spots since I made it. As the body's protective covering, skin bears the brunt of assaults from a rapacious outside world constantly tearing at our bodies, as well as from the sun's harsh rays shining on it. It seems fitting that animal skin can become clothing for humans, an extra layer of protection added atop our own biological one to further fend off the world's aggressions. By harvesting and softening the skins of animals, ancient humanity learned to create yet another layer—this one artificial—atop the many layers of their own skin.

Skin tells a story about the layers from which our bodies are made, but it also tells another story—one about how humankind creates its own material world outside the body. After clothing, people learned to craft additional layers outside the body, including the walls of our houses, an additional protective covering that gives the body a small bubble of safety in a dangerous world. And we build walls with the same layered pattern as our skins—a sturdy middle layer of wooden studs or concrete gives hidden strength like dermis, while an outer layer of superficial siding sheds water like epidermis and becomes the only layer visible from the outside. The same histological blueprint of our own bodies is the model we use to craft protection from the elements, always building in our own biological image.

9

URINE

Most people do not have a favorite bodily fluid—they are all generally considered repulsive. Even in the medical field, there is an old adage that physicians choose their careers based on the bodily fluid they find least revolting. Since each medical specialty has its own essential fluid, a doctor disgusted by stool and sputum, but able to stand the sight of blood, might end up a hematologist, while one repulsed by urine and bile, but tolerant of sputum, might gravitate toward the field of pulmonology.

As a medical student, however, I preferred to think that some physicians were actively *drawn* to a particular bodily fluid, intrigued by its unique diagnostic mysteries. From the pus of infectious disease specialists to the nasal drainage of ear, nose, and throat surgeons, all of the body's many excretions, secretions, and suppurations are an essential source of information used by physicians to diagnose and, ultimately, alleviate maladies. Bodily fluids, which are typically discarded and disdained, are the medium of a physician's craft, and each has its own language in which it speaks to physicians, telling them what is wrong with a patient. Being a specialist means becoming fluent in one specific fluid's dialect, learning to interpret its colors, textures, and consistencies, and spending a career pondering its secrets.

Choosing a medical career felt like choosing a bodily fluid. And though I resisted settling on just one (I remain a generalist), I have always been partial to urine.

Conceived in the kidneys—a pair of bean-shaped organs tucked away in the abdomen's rear—urine consists mostly of water filtered out of the bloodstream, along with the body's liquid waste products, which contribute to its color and smell. Urine runs from the kidneys down the ureters and is conveniently stored in the bladder, from which it is needle-lessly gathered by urinating into plastic cups for testing. Urine stands out by the wealth of information it grants doctors about a patient's condition, and urine analysis is performed frequently enough by physicians to have earned the shorthand *urinalysis*. No other bodily fluid can claim to be on a nickname basis with the medical profession.

I remember the first time I watched a nephrologist—a specialist in kidneys—turn a urine sample into a diagnosis. During my nephrology elective as a medical student, I followed behind as he carried a small plastic urine cup to the microscope room in the nephrology department, his long white coat flapping behind him as he moved swiftly through the hospital's halls. He plunged a diagnostic dipstick into the fluid, which revealed bits of blood and protein unseen by the naked eye. He then used a plastic pipette to place some urine into a centrifuge, which spun rapidly and concentrated floating cells into a sediment at the bottom of the vial. After peering through a microscope at a single drop of this stuff, noting stray bits of debris flung across the viewing field, he wove a comprehensive diagnostic tale that encompassed all of the patient's symptoms and lab abnormalities. The diagnosis turned out to be glomerulonephritis, a rare form of kidney disease. He was able to see inside that patient with a clairvoyance that seemed positively sorcerous, with urine as the crystal ball. From that moment, I was determined to learn urine's secret language.

In the months that followed, I practiced wielding the dipstick and centrifuge, and trained my eyes to recognize clues under the microscope. I befriended technicians in the hospital lab and had them save interesting urine specimens that might broaden my experience of disease. A designated cup with my name on it soon appeared in the laboratory fridge, and each day I stopped by during lunch to examine their latest finds.

Squinting through the microscope lens one day, I spied budding yeast in the urine of a critically ill patient with a fungal infection in his bloodstream. In what proved to be a fatal condition, the fungus had overwhelmed the kidney's filters and leaked through into the urine. On another day, I saw sexually transmitted parasites—the teardrop-shaped creatures called trichomonas eagerly swam and pirouetted like water ballerinas in the man's urine. His specimen told of unprotected sex, and I wondered with how many other unwitting partners he had shared his infection. My nascent knowledge of trichomonas infection told me that he may have sought medical help because of burning with urination, or perhaps, as is also common, he had no symptoms at all. Through urine I felt as though I could get to know the lives of people I would never actually meet simply by reading their disembodied effluent as evidence. Each new discovery was a thrill that gave me the feeling of disparate threads suddenly connecting together—the same feeling, I imagined, a detective feels when stumbling on a keystone clue.

When my own patients needed a urinalysis, I performed it myself rather than sending samples to the laboratory only to receive back computerized results. With my own (gloved) hands, I carried the patients' still-warm samples down the hall to the microscope lab and personally extracted their clues. Doing it myself gave me a deeper sense of the pathologies playing out inside my patients, and repeating the ritual of dipstick and centrifuge again and again created an intimacy with them—and a growing respect for my favorite bodily fluid.

I quickly began to comprehend urine's enigmatic language: white blood cells on the dipstick spoke of urinary tract infection, while crystals under the microscope might suggest kidney stones and explain some cases of flank pain. Each analytical finding became an important diagnostic message communicated in the language of urine.

Of course, urine is not the only bodily fluid that allows a clinician to indirectly peer inside the human body and diagnose its ailments. Stool brings a clinician information about the entire gastrointestinal tract from mouth and nose to colon; sputum delivers news from deep within the respiratory tree; and cerebrospinal fluid tells tales of an inaccessible central nervous system. Just like urine, these other bodily discharges sweep evidence of disease out to a waiting physician, a more primitive and anticipatory version of using diagnostic scopes to enter the body's orifices.

But more than any other fluids, urine gives information about not only troubles within its own exit route through the urinary tract but also the body as a whole—even systemic, seemingly unrelated problems. Red urine can imply problems like kidney stones or bladder cancer, but it can also indicate a genetic defect in red blood cells, a breakdown of the body's muscles, or a recent meal of beets. Urine can uncover a cause of infection in the faraway lungs, announce a recent ingestion of recreational drugs, or, in diabetes, confess the failings of a distant pancreas. In previous centuries, physicians made this last diagnosis by tasting urine for sweetness, a time when taste buds were the only assay. Today, thankfully, this is no longer necessary, though decoding all of urine's secrets still often feels like being a sommelier.

Years after medical school, I worked in an ER and discovered an even more fundamental reason to appreciate urine. In questioning every ER patient that I saw, I always brought the conversation back to urine, no matter the primary problem. Whether the medical issue was fever, vomiting, diarrhea, or cough, I invariably asked how much a patient had urinated lately. Though a measure far more elementary than that requiring a dipstick or microscope, the simple volume of urine produced by the kidneys is an essential indicator of a person's hydration level.

When the body becomes dehydrated, through poor intake or fluids lost to vomiting and diarrhea, the normally surging, turgid bloodstream becomes sluggish, like a dried-up stream in a drought, and the delivery of nutrients to vital organs diminishes. The kidneys perceive slackening in the blood vessels and respond by slowing or turning off the urine faucet, preserving the body's hydration while letting out only minuscule amounts of dark yellow and sometimes malodorous urine. When a patient's urine output dwindled to a dangerously low level because of disease or infection, it raised a red flag for more serious illness, and such patients often needed further diagnostics and intravenous hydration. This was urine's crucial message in the ill.

The standard treatment for dehydration is intravenous salt water, or saline. In those with infections ranging from mild to severe, intravenous saline often makes them feel better and brings alarming vital signs back to a normal range. It also causes a patient's urine stream to pick up, turning from a slowed yellow trickle into a clear watery deluge—another message from urine, this time one of clinical improvement. More than any wonder drug or miracle of modern medicine, a simple salt solution is the most common treatment I infuse into my patients' veins.

But as a first-year medical student, I often wondered why the solutions doctors infuse through patients' IVs were always salty. After all, human life seems dependent on fresh water with virtually no salt in it. Fresh water is the only kind of fluid that quenches our thirst, waters our crops, and bathes our bodies. But if I infused the same fresh, saltless water into my patients' veins, it would be toxic, quickly causing blood cells to burst and the brain to swell, leading to seizure, coma, and death. Normal saline, the IV fluid most commonly given, is a standardized solution with the sodium chloride content of a heavily salted soup. And it highlighted the fact that, though we seem to live in a freshwater world, only salt water can truly fill our hearts. But why?

I found the answer to my question in urine and the way kidneys make it. By finely calibrating the urine extracted from the bloodstream at every moment of our lives, kidneys safeguard not only the body's hydration but also its salts. Though mostly water, urine carries salts as well, thereby maintaining the body's essential balance of water and electrolytes.

In illness, the kidneys work overtime to accomplish this task, struggling in the face of dehydration to keep blood high in sodium and chloride and low in potassium—roughly the same salt proportions as the world's oceans. Humanity's ancestors first evolved in the sea and subsequently crawled out to take up residence on dry land, but each human body still carries the ocean inside it. By editing urine out of the bloodstream, kidneys preserve the primordial sea in our blood, maintaining the balance of salt essential to our survival. Without them, and without urine's salubrious flow, our forebears could never have left the ocean, just as each newborn baby could never adjust to life outside its personal salty amniotic sea—itself composed almost wholly from the unborn baby's urine. So, when urine flow slows in illness (when patients report poor urine output or parents tell of fewer wet diapers in sick infants), it is the body fighting to maintain the life-giving ocean inside each of us, our ancestral brine.

My love of urine led to a profound respect for the kidneys. They are among the most metabolically active organs in the body because of their constant titrating of blood and equilibrating of urine, each kidney an expert chef after just the right mix of ingredients. Despite their overarching importance to the body's biochemical balances, kidneys churn away in obscurity our entire lives. They are relegated to the back of the abdomen, hidden behind more charismatic and well-known abdominal organs. Kidneys are the Rodney Dangerfield of our innards—they never get the respect they deserve.

Yet the careful harmony protected by the kidneys' lifelong labor of making urine ensures that other bodily systems also continue to function. This includes the electrical activity responsible for every heartbeat, every muscle movement, and every neuron firing in the brain, each a manifestation of the ebb and flow of salt ions through water. It is precisely because the kidneys receive blood and turn it into urine that the brain can receive blood and turn it into thought.

As a child, I sang about the knee bone's connection to the thigh bone, but in becoming a physician, I discovered much deeper connections among our body parts. I came to appreciate the ecology of our internal organs, and it was through watching my patients' kidneys—specifically by watching them fail in their urine-making duties—that this concept became a reality.

My first close-up experience with kidney failure came while working in the ICU as a resident, caring for patients closest to death. Each morning, I rounded through the ICU's brightly lit halls as part of a massive team of physicians, nurses, therapists, and a pharmacist or two. We reviewed each patient's condition and lab values, and we numbered their failing organs. I watched organs of all sorts fail in the ICU: when the heart failed, blood would back up and fill a patient's body with fluid; when the lungs failed, patients needed a ventilator to breathe for them; when intestines failed, food's transit became stalled, leading to bloating and vomiting; and when the liver failed, patients would become jaundiced and distended with abdominal fluid. I even saw what could be called "brain failure," as severe illness frequently left my patients in and out of consciousness, disoriented, delirious, and sometimes frankly psychotic—the ICU has the most severe cases of every disease, including hospital-acquired delirium.

I was surprised to find that the kidneys failed more than any other organ. When this happened, the flow of urine often slowed significantly, sometimes sputtering to a complete halt. In other patients with kidney failure, urine production continued but its salt content became altered, resulting in dangerous salt imbalances and a buildup of waste in my patients' bloodstreams. This happened often, even in patients suffering from diseases completely unrelated to the urinary tract. The body's urine-makers seemed uniquely fragile in the face of almost any kind of critical illness.

I saw this pattern of the kidneys going awry most prominently in my ICU patients with liver failure. On morning rounds, as the team recited each patient's daily deluge of blood and urine test results, the patients with severe liver failure often suddenly began showing signs of kidney failure too. When we discovered a drop in their urine output or a buildup of waste and certain electrolytes in their bloodstreams, we conducted the usual battery of new tests to figure out why the kidneys had joined the liver in failure. But we often found no problem within the kidney itself.

To be sure, more than one organ failing at a time in an ICU patient was nothing out of the ordinary, and the physiologic explanation for why any two organs failed in concert was easy to understand. In patients suffering from heart failure, fluid backup flooded their lungs, causing difficulty breathing and, ultimately, lung failure too. Patients with liver failure often developed brain failure, or delirium, as unmetabolized toxins disrupted neuronal function. And brain failure—whether temporary delirium or permanent dementia—caused lung failure when a patient aspirated.

With failure of the kidneys and liver, however, there was no good explanation. Though these two organs share a habitat—the right kidney pressing intimately up against the liver's backside—they seem to be disconnected from each other in nearly every other way. They have markedly different functions and have no specific arteries or veins in common. The liver and kidneys are as physiologically distant from each other as two organs under the same abdominal roof can be. Yet in the ICU, I was repeatedly faced with their synchronized failings.

The condition has its own name: hepatorenal syndrome. This vague appellation simply refers to both liver (*hepato*) and kidneys (*renal*), while the word *syndrome* is Latin for "running together," and it signifies a collection of signs and symptoms often predictably found together but without a clear unifying principle. *Syndrome* is code for the medical profession's ignorance about why the human body acts in certain ways. In the case of hepatorenal syndrome, we still do not understand the underlying interdependence of kidneys and liver, even though we know that kidney failure is often a harbinger of a cirrhotic patient's rapidly approaching death. And it remains equally mysterious why, when the kidneys have all but shut down in these patients, a liver transplant alone often magically restores kidney function.

Just as I learned to love urine because of its ability to speak to me about disparate aspects of the body, I learned that kidneys, too, had their own clairvoyance. In patients with liver failure, I could monitor kidney function as an indirect way of keeping an eye on the liver. Though I could not adequately explain the genesis of the syndrome, I could anticipate it. And the kidneys provided a red flag in other types of severe illness too. *Cardiorenal syndrome* denotes end-stage heart failure leading to kidney failure for poorly understood reasons, and, as with the liver, once the flow of urine is altered and kidneys begin to fail, it often means the end is near. *Multiple organ dysfunction syndrome* (MODS) refers to a catastrophic cascade of organ failure like dominoes in the critically ill, the human body's version of ecosystem collapse. And kidneys are often the first to go, followed by a chain reaction that leads inexorably to death.

Diagnosing and managing hepatorenal syndrome showed me the importance of seeing my patients ecologically. Organs show their interconnectedness in health, but they reveal even greater mutual dependence in illness. For this reason, physicians often make diagnoses indirectly. We sometimes discover pancreatic cancer only once it affects the liver and biliary system, and thereby comes to medical attention, while ovarian cancer often manifests only once it disturbs the bowel. In the same way, critical disease anywhere within my ICU patients could speak to me through the flow and content of my patient's urine. Human bodies are more than just a collection of individual parts, and clues to the well-being of one organ lie hidden within another.

Before going to medical school, I understood ecology in nature's ecosystems, but in training to be a physician, I saw the essence of medicine to be a similar ecological understanding of our insides. Just like the natural world, the body contains many symbiotic relationships, and most make perfect sense. The heart and lungs, for instance, share a thoracic habitat, and they speed up in sync when a person runs up a flight of stairs, and then slow back down together when that person rests, rising and falling just as the population of a predator and its prey rise and fall together in a balanced harmony. Other symbiotic relationships, however, such as the kidneys and liver, are less obvious, like the mycorrhizal connections between tree roots and fungi hidden beneath the ground. Our interconnected innards communicate with each other through a subtle and complex language of bloodstream signals that medical science has not fully decoded. Like wolves howling in the night and sending cryptic messages drifting through the air, blood-borne directives carry meanings at which we can only guess.

In nature, the impact of one species on another is often impossible to predict with our limited knowledge, which is why the extinction of any species can threaten the ecological balance in unanticipated ways. And, as with natural ecosystems, the breakdown in any one part our internal ecology can lead to failure of the global order, even if we cannot explain how with the most advanced scientific understandings.

The natural world and its varied inhabitants, along with other members of our own species, surround our bodies like an outline, a cast of interconnected characters filling the lives we live on the outside of our bodies. But inside our bodies is another set of characters composing an internal ecosystem that echoes the natural world. Each of us fits into a larger whole, and we each contain within us a similarly unified and interwoven totality—an ecology within an ecology.

Kidneys may not occupy prime real estate in our bodies, but through their production of urine, they assist and support all other body parts, a keystone species in the body's internal ecosystem. Without the continuous flow of vital urine, the entire system comes crashing down. Instead of being fragile, kidneys are prescient—a canary in the body's coal mine. And urine is the language in which they scream for help.

10

FAT

was walking down a street in Barrow, Alaska, when I turned a corner and saw it—piles and piles of glistening whale blubber strewn over the snowy lawn in front of a small green trailer. Each strip was about as long as a person, two feet wide, and nearly a foot thick, and the fat was speckled in shades of pink and yellow tie-dye. I poked at one of the strips with my finger—it was firm, oily, and cool to the touch. Scattered among the mounds of fat was the rest of the bowhead whale: there were chunks of meat; black strips of baleen, which the whale had used to filter its food out of seawater; and a massive heart as big as a yoga ball with an attached stump of aorta the diameter of my thigh. Off to one side sat a wooden sled, which had been used to ferry all this flesh from out on the sea ice. And just beside it lay a long knife blade, the snow beneath it stained red with blood.

I'd come to Alaska to work in Anchorage on a public health research elective studying infection risk among Alaskan Natives. I had always been fascinated by the impact of extreme conditions on the human body, and during a break from my research, I decided to visit Barrow, the northernmost town in the United States. What I found there was a lesson in how a harsh geography and climate molded a diet unique in the world.

As I gawked at the blubber, the trailer's front door opened and a man in a baseball cap and dark sunglasses walked out, a large metal hook in his right hand. He walked toward a piece of fat about the size of a football lying on the ground. Straddling it, he drove the hook's sharp point into the chunk, which jiggled from the impact. He walked back inside with the blubber dangling from the end of his hook as smoke poured from the metal chimney above the house. For the people inside that green trailer, all this blubber was food, and it meant sustenance for months to come.

Before the Western world "discovered" Arctic Alaska, the Iñupiat people in this region lived completely off the land and sea. The region has an overwhelming abundance of blubber-laden animals like whales, walruses, and seals—but virtually no fruits, vegetables, or grains—and the traditional Iñupiaq diet reflected these realties. Virtually all food was animal based, and like the Atkins or paleo diet, it was low in carbohydrates and extremely high in protein and fat; animal fat made up more than half of all the calories that people consumed. The same fat that served marine mammals as energy storage on their bodies during life filled an Iñupiat family's underground ice cellar after death in another kind of life-preserving energy repository co-opted from the natural world by humans.

The traditional Iñupiaq diet goes against every dogma I learned about health and nutrition in medical school. As a physician, I was trained to steer my patients away from diets high in fat, particularly animal fat, in order to reduce their risk of heart attacks, strokes, and metabolic disease. I was also taught to warn my patients about obesity in the same breath, though I left medical school with only a muddled understanding of precisely how fat in my patients' diets related to the fat packed onto their bodies in obesity, or how either related to fat in their bloodstreams in the form of cholesterol and triglycerides. The latest in nutrition science offered only confusing and contradictory explanations with little hard evidence to support the main tenets of what I advised my patients. Just one thing was clear: fat in all its manifestations was the enemy.

In the Arctic, however, fat has always meant health and survival. Even though people in Barrow no longer depend exclusively on the land and sea to eat, a lawn covered with blubber is still the equivalent of a plush, green lawn in temperate suburbia swelling its owner with pride. I was used to fat as the most maligned of all body parts and the culprit in an obesity epidemic, but here fat was celebrated. The story of life in the Arctic, especially human life, is basically a tale of fat.

My cadaver in anatomy lab had died an obese man. Once we began the dissection, the first thing I saw beneath his skin was a two-inch layer of fat across his lower back. It was translucent with the yellowed tint of aged cheese, and it made a crunching sound as we sliced through it with repeated scalpel thrusts, working past it toward the back muscles that were the subject of the first day's lesson.

Fat is best known for lying just beneath the skin, but I found a lot more of it deep inside my cadaver's body, swaddling his organs. His heart was wrapped in yellow adipose, and it had a color scheme that was the complete opposite of the photo of a heart in my medical textbook—instead of a meaty red organ with yellow highlights, his heart was almost all yellow, with just a few hints of red muscle showing through. His intestines were blanketed by greasy yellow swaths, and small blobs of fat dangled from the length of his colon like chunky jewelry. Once we moved the intestines aside and looked into the abdominal recesses where his kidneys should have been, we found two large yellow blobs. Only when we sliced them open did we find his kidneys.

My cadaver's fat proved to be a nuisance. It left a greasy residue on my gloves and metal dissecting instruments, making them slippery and more difficult to handle. Tiny fat globs adhered to everything, and the green scrubs I wore quickly became festooned with dark grease stains. The temperature in the anatomy lab was kept cool to reduce smells and keep fat firm, but neither goal was achieved—my cadaver's fat smelled like chicken soup and coated everything, making each step of the dissection more complicated.

When I decided on that first day to donate my body for a medical school dissection, I thought the decision might have an unintended side benefit: it would motivate me to stay slim as I aged. The average American adult gains a few pounds every year, meaning fat increasingly accumulates on the body. For the sake of the medical students who would be presiding over my greasy dissection, if not also for myself, I hoped that I could buck the average trend.

I thought about the life my cadaver had lived, and wondered how his copious body stores of fat had weighed him down in life, a strain on his internal organs as well as his knees. Perhaps he also suffered from the stigma that many obese people experience in all aspects of life, especially when accessing healthcare. At the very least, he would have been on the receiving end of much confusing and contradictory nutrition advice from his doctor.

Once I started caring for living bodies in the hospital, I found it more difficult to examine obese patients and to make a diagnosis. A layer of fat made the sounds of heart and lungs more distant through a stethoscope, and my inability to delineate abdominal organs with my hands flustered me. Reading my patients' neck veins to characterize problems in the heart was an important physical exam skill that required repetition and practice to master, but in most of my patients—even those only slightly overweight—I could barely make out the subtle vein pulsations because of neck fat. I took the opportunity to practice this skill whenever I treated cancer patients wasting away from their disease, since their neck veins were readily visible. I never achieved proficiency, but it did not seem to matter all that much, since the attendings usually ordered X-rays and other imaging studies that provided similar information about my patients.

When I traveled to Mumbai as a medical student, I witnessed the practice of medicine devoid of obesity's hindrances. The mostly poor Indian patients seeking help at the public hospital where I worked were universally skinny, the result of poverty, poor nutrition, and the fact that they could not afford to seek medical care until they were extremely ill and wasting away from their disease. But in such a thin population, the physical diagnosis maneuvers that I had learned seemed easier and more useful than they ever did in the United States.

I was deeply impressed by the examination skills of the Indian doctors—my American attendings seemed inept in comparison. It helped that the poor Indian patients they were treating could not pay for expensive imaging tests, so doctors were forced to rely on their physical exam alone to make a diagnosis. While American physicians are rightly faulted for overusing imaging technologies like CT and MRI scans, such studies are more necessary when the utility of the physical exam diminishes in very overweight patients—and the less American physicians rely on hands-on exam techniques, the more skills are lost, and a vicious cycle quickens.

Once I managed to make a diagnosis in my obese patients, fat got in the way of delivering treatments too. I struggled to locate their arm veins, and found it more challenging to thread them with IVs. When multiple attempts were needed, it caused more pain for patients and frustrating delays in urgent treatment. Inserting a breathing tube into a patient could be lifesaving, but with obese patients, it was more difficult—reduced mobility in the neck and a crowded pharynx made it harder to get a good view of the windpipe. Failing to get the tube in the right place can mean death, which is why the thought of needing to intubate very obese patients can make doctors sweat. And spinal taps were almost impossible on these same patients—a thick layer of fat over a patient's lower back, like my cadaver's, made attempts to precisely target the spinal canal with a cartoonishly long needle seem utterly futile.

Fat made almost every aspect of practicing medicine more difficult, and my obese patients needlessly suffered.

Spring in Barrow is whaling season, and I wanted to learn more about how the Iñupiat people experience fat. I asked around and met Herman Ahsoak, a whaling captain who was heading out onto the sea ice to hunt whales the following day. I asked if I could tag along, and once I convinced him that I did not work for Greenpeace—an organization that has sometimes battled against the traditional whale hunt—he agreed to let me join his whaling crew.

The following morning, we sped away from shore on snowmobiles over the frozen Arctic Ocean, our crew consisting of Herman, his teenage son and daughter, and his friend Greg. As we moved farther from shore, Barrow slowly sank below the horizon behind us, leaving only ice and sky in all directions. The frozen landscape of pure, unblemished, and unbearably bright white was finally broken after some miles by a dark streak of open water. We set up camp along the ice's edge and waited for bowhead whales to appear in the water.

And we waited. Most hunting is similar to being a pediatrician at a birth: it consists of long boring stretches of inactivity punctuated by a sudden explosion of excitement and action that can arise at any moment. I vigilantly kept an eye on the open water, watching for whale blows in the distance and admiring blue icebergs floating by in a leisurely current. The water looked so cold that even the most obese human would not stand much of a chance if they fell in—fat in whatever amount is not distributed on the human body as properly as it is on marine mammal bodies in need of insulation.

Unlike the whaling industry of the early twentieth century, which used massive floating factories to slaughter and process whales in huge numbers, and which decimated whale populations throughout the Arctic, Herman hunts in the traditional way. He had dragged behind his snowmobile an umiak, a traditional boat made with a wooden frame and covered by a hull of seal skins sewn together. The umiak was placed at the ice's edge with its snout sticking out over the water, ready to be launched at a moment's notice if a bowhead whale came close enough. Inside the boat lay the exploding harpoon that Greg had brought—a modern, and more effective, version of the stone- and ivory-tipped weapons his ancestors would have used.

Right beside our camp, a trail of fresh polar bear paw prints braided along the ice edge. The large prints framed by the feathery brushings of shaggy fur unnerved me—as often as I looked at the open water for whales, I looked behind me for bears peeking out from behind the white ice chunks littering the landscape. Herman guessed the bear was hunting seals, and, he added, they always eat the blubber first. Like the Iñupiat people, polar bears know to go for the money, the Arctic's universal currency of fat. The thought of a bear lurking nearby made me conscious of my own belly flab and love handles in a new way—as a staple food.

I asked Herman about the importance of fat in Iñupiaq culture, and he pulled out a plastic bag and unwrapped from it a brick of raw whale blubber. The brick was white and capped with an inch-thick strip of black skin, a traditional Iñupiaq favorite called muktuk. He cut thin slices with a pocketknife and passed them around. I bit into one and began chewing, my lips coated with an oily slick. It tasted like the sea smells, and it was not nearly as tough as I had expected.

As we snacked, Greg, who is half Iñupiaq, complained that people often doubt his Native ethnicity. His skin was whiter than Herman's, though he spoke with the same accent. As definitive evidence of his heritage, he professed his love of muktuk.

"I may look white, but I have an Iñupiaq soul," he insisted, as though eating fat (especially raw) were the quintessence of Iñupiaq identity.

Later that day, Herman brought out a large white plastic bucket that was three-quarters full of what looked like pure oil and smelled like a barnyard. He plunged a wooden spoon into the bucket and stirred until a few dark black strips appeared floating to the top. He called this dish kiniqtaq— seal jerky marinated in rendered seal blubber. The piece he handed me was completely saturated in fat and dripping large oily plops back into the bucket. It was chewy, greasy, and pungent, but good. I enjoyed a second and a third piece while Herman's teenage children preferred to snack on chips purchased from Barrow's supermarket.

The same seal oil filling the bucket, Herman explained, is an all-purpose condiment in Iñupiaq cuisine: a dipping sauce, a flavor bomb, and a cooking oil. It also used to be the primary fuel for lamps that the Iñupiat traditionally fashioned out of stone. Since the region has virtually no trees and, therefore, little wood to burn, animal fat was the only combustible material abundant enough to provide light and heat. Burning fat in a lamp was the biochemical equivalent of eating and metabolizing it within the body to produce heat and energy, and Herman and Greg insisted that no other kind of food besides fat could keep them warm while hunting out on the exposed ice.

Fat has been the keystone of human survival in the Arctic, and the Iñupiat often risked their lives to harvest it. I thought about the ice beneath us—a few measly feet of frost already in the process of melting held us above certain hypothermic death. Tragedies were frequent throughout history. Greg talked about the time he floated away on a piece of ice along with seventy other people. The piece of ice was so large that they did not even realize they were floating away until the news came over the shortwave radio. Each person, and every piece of equipment, was saved in a helicopter rescue operation, though, in the past, all those people would likely never have been heard from again. To think, the Iñupiat people once traveled onto this fickle spring ice with dogsleds just to put blubber on the table.

Around lunchtime on our second day hunting, we finally saw whales. A puff of water appeared far away in the open water, and suddenly a whole pod of bowheads was jumping around and slapping the water with their tails and flippers in a chorus. I stood up and shouted with amazement, and Herman quickly shushed me and motioned for me to squat back down. I crouched behind a piece of ice and stared—it was an impressive display of acrobatics for such leviathans. Their black skin looked perfectly smooth and shone brilliantly in the sunlight, and just beneath it, out of sight, was a layer of protective and nutritious blubber.

The bowheads swam away without ever coming close enough for us to launch an attack. We stayed out on the ice for another day, and the innumerable seals and ducks passing by showcased the explosion of life that is summer in the Arctic. But we did not see another whale, and we returned home from the ice empty-handed.

Obesity is a fact of modern life, and it has become widespread in Arctic Alaska, just as it has everywhere else in the country. Herman was one of the few slim people I met in the region. Of all native peoples contacted by

Western society and yanked into the modern world, the Iñupiat had among the most abrupt trips. In just a few generations, they went from a paleolithic existence, and a paleo diet, to a typical sedentary modern American existence with motorized transport and supermarkets full of food. Carbohydrates and processed foods intruded on their carnivorous, blubber-filled diet. However, at the same time that fat was becoming less of a food staple, and also disappearing from oil lamps, it increasingly appeared on their bodies. This transition directly into the modern obesity epidemic was greatly assisted by significantly less exercise—Iñupiat people went from living an incredibly rigorous subsistence lifestyle to being couch potatoes like the rest of us, a process of modernization in fast-forward.

In the past, the Iñupiat, like most people living off the land, built up fat stores on their bodies during seasons of plenty, and it helped them survive through times of want, which were common. Fat on the human body is another future-oriented organ, like genitals, an insurance against poor hunting seasons to come. Even fat's appearance under a microscope as bubbles packed tightly together with hardly any room to spare suggests its function as pure storage space for the body. But in the modern Arctic, like most other places in the country, there is no longer feast and famine—it is only feast.

Two years after my trip to Barrow, I worked in a hospital in another part of Arctic Alaska, where I saw extremely high rates of obesity. As a man who had relocated to the Far North from the continental United States remarked to me, "Everyone puts on weight when they move up here." Just as migrating birds and sea mammals like bowhead whales grow fat in the Arctic, so do humans.

To better understand the confusing relationship between fat in our diets and fat on and in our bodies, I spoke with Dr. Lee Kaplan, director of the Obesity, Metabolism & Nutrition Institute at MGH in Boston.

His answer to the puzzle of fat and health: "There is no single answer."

Kaplan explained that not all fat on the body is unhealthy, nor is all fat in the diet, and the relationship of either to metabolic diseases like high blood pressure, high cholesterol, and diabetes is even less clear. Genetic differences across ethnicities make a one-size-fits-all solution to obesity impossible. He pointed to a recent study showing that specific genetic adaptations in natives of the Far North help their bodies use omega-3 fatty acids, a type of fat abundant in blubber, more effectively than other populations. This helps explain why the Iñupiat could historically be healthy and have very low levels of cholesterol despite eating diets rich in animal fat.

Even within the same ethnic group, Kaplan explained, there is tremendous genetic variation from person to person. Every individual human body has its own set point for the amount of fat it carries—Kaplan compared it to the temperature setting on a house's thermostat. A body clings to that weight regardless of diet or exercise regimen, and as a result, some skinny people can have as much trouble gaining weight as obese people trying to lose it. Yet the fat on our bodies defies any single physiologic understanding: the kind wrapping around our internal organs acts very differently physiologically from the kind under our skin. Fat is actually multiple different organs, and science is only beginning to understand their many nuances.

Kaplan reserved his harshest words for doctors themselves: "We call obesity a disease, but we treat it differently than all other diseases." For all the other ailments caused by a modern lifestyle, he said, "we just have to find an effective drug." He pointed to the wide variety of different medications available to treat high blood pressure, high cholesterol, and diabetes. But obesity is different. He explained, "When it comes to obesity, we say that the only way to solve the epidemic is to completely change modern life." He offered a theory as to why doctors see this condition differently: "People wear obesity on the outside of their bodies, and you can regulate your weight in the short term by starving yourself." As a result, doctors tend to see it as a personal failing rather than a disease, and Kaplan believes that society as a whole has followed the lead of the medical community and adopted that same attitude.

I immediately thought of another condition that has elements of both medical disease and a personal lack of control—addiction. In the context of our current opioid epidemic, physicians are slowly beginning to view addiction as a disease, rather than a personal failing or a crime, and finally starting to treat it with medications that have proven effective. I asked Kaplan about a few new medications to treat obesity that had come on the market in recent years; he said they were "woefully underused by doctors in the US." He blamed ignorance about the causes of obesity and admonished me and my fellow physicians: "Be consistent."

The truth is that we're only beginning to unravel the mysteries of diet, fat, and disease. The medical community's ignorance, as well as our biases, means that the nutrition advice we give to patients changes constantly. Each decade of the last half century found a new food to demonize, and soon after, the recommendations reversed themselves. "The result," said Kaplan, "is that the public thinks we're idiots."

When Kaplan evaluates patients in his institute, he not only focuses on diet and exercise but also investigates how stressed a person is, how well they are sleeping, and their exposure to certain pollutants, all factors that appear to raise the human body's fat set point and lead to weight gain. The simple math equation of calories in and calories out is a gross simplification of human metabolism and unhelpful for physicians advising their patients. Eat less and exercise more is a sure way to alienate obese patients, since the condition is almost never that simple. As Kaplan said, "Five thousand out of a total of twenty-two thousand genes in our bodies are involved in metabolism. Why wouldn't it be complicated? There is almost an infinite amount of complexity."

Several years into my career in medicine, I treated the heaviest patient I had ever seen. Natalie was a forty-eight-year-old morbidly obese woman who weighed over five hundred pounds, and she came into the ER because of severe pain in her right leg. I walked into her room and was immediately struck by her bulk—she easily filled the large hospital bed and practically spilled over both sides of it. She had streaks of gray in her blond hair and a contorted look of anguish on her face.

I could see that her right leg was mottled purple and white. When I placed my hand on her foot, it felt ice cold, and I immediately knew that blood flow to her leg was being blocked. I searched the top of her foot for a pulse, but I could not find it, either with my fingers or with the Doppler machine the nurse had wheeled in.

She needed an urgent CT scan to figure out what had happened, but I also wanted to get her some pain relief as soon as possible. I asked the nurse to put in an IV and give her a dose of morphine—twice the usual number of milligrams owing to her size. Shortly after I ordered a CT scan, a technician from the radiology department called to tell me that Natalie was too big for the scanner: she was well above the machine's weight limit.

I canceled the CT order and instead ordered an ultrasound of the blood vessels in her leg, which confirmed that no blood was flowing through her arteries. The ultrasound could not see deeply enough inside her to tell me exactly what was responsible for obstructing the blood flow, though I suspected either a blood clot or a torn artery. Regardless of the cause, Natalie likely needed vascular surgery to relieve the blockage, but the nearest vascular surgeon was at least an hour away, and time was of the essence. Her leg could survive for about six hours from the time blood flow first stopped, and it had already been three when she arrived in the ER. If I did not get her transferred quickly, she would likely lose her leg.

I called the closest large hospital and spoke with the vascular surgeon on call, but she refused to accept the transfer. After checking with her hospital's own CT technician, she learned that Natalie was too heavy for their CT scanner, too, and they wouldn't be able to do the surgery anyway, because she wouldn't fit on their operating room beds. I hung up, annoyed.

While I contemplated my next move, the nurse told me that Natalie was still in severe pain, so I ordered another shot of morphine in the same large amount. I called a second hospital, and a third, and got the same runaround—the same canned rejection. Transferring patients always makes me feel like a salesman trying to convince prospective buyers, and it often demands I spend too much time on the phone; with Natalie, I was failing to make a sale, and one call after another was pulling me away from other patients who needed my attention. A feeling of dread materialized in the pit of my stomach, and after being refused by a fourth hospital, the dread grew into panic. I yelled rudely at the fourth vascular surgeon as I hung up the phone. In their minds, I imagined, Natalie was simply too fat to care about.

Meanwhile, the morphine had hardly touched her pain—I ordered more, and more, nervous that each dose would suppress her breathing further. With each injection of painkillers, I worried that giving her too much could lead to respiratory failure, and her size would make her almost impossible to intubate. But I also didn't want her to suffer.

Finally, the fifth hospital accepted the transfer—but it took another hour to locate an ambulance with the appropriate bariatric stretcher to transport her. The six-hour window had passed, and I felt defeated. To top it all off, I had hardly even managed to dull her pain, despite giving her more morphine than I have ever given any human in such a short amount of time.

Three days later, I spoke with the vascular surgeon who had taken Natalie under his care. He had confirmed my suspicions and discovered a blood clot in the leg artery, and though he had removed it, Natalie's leg was already too far gone to save. On the next day, the same thing happened to Natalie's other leg: it became cold, painful, and mottled in color, and the surgeon had found a second blood clot, which he was able to unclog in time.

But on Natalie's third day in the hospital, a third blood clot appeared— this one a pulmonary embolism that lodged in her lungs. Despite all efforts, her heart went into cardiac arrest, and she could not be revived. The surgeon and I shared our surprise and dismay at this terrible outcome, and we both acknowledged our befuddlement at such exorbitant clotting. The surgeon mentioned that Natalie's obesity put her at risk, and I concurred—fat acts like an endocrine organ and secretes hormones that can promote clot formation. It hurt me even more knowing that I had failed to dull Natalie's horrific pain in what would turn out to be one of her last days of life. Whenever a patient inexplicably dies, especially at a young age, I go over the case again and again in my head for days, and sometimes for years, wondering what I could have done differently—and Natalie's case is one on which I will likely never stop ruminating.

While fat on the human body has repercussions for the quality and speed of healthcare delivery, my experience with Natalie highlighted systemic and technical barriers to delivering medical care to morbidly obese patients. Though she might have died no matter what I did, Natalie's weight contributed acutely to her pain and suffering, while showing the limitations that are inherent in every piece of medical equipment, from crutches to wheelchairs to CT scanners.

A few months after Natalie's death, I treated an even more morbidly obese man who weighed over eight hundred pounds. He had lived in a nursing home for years because of the disability caused by his weight, and when he developed a cough and fever and needed to go to the hospital for evaluation, it took more than fifty firemen seven hours to extract him from the nursing home. They removed a large bay window and constructed a ramp from scratch so that his bed could be wheeled through it and into a box truck, which is how he was transported to the hospital.

The obese are often dehumanized when encountering the healthcare system, and a trip in a box truck represents only the most extreme example. Many people with mild or moderate obesity encounter innumerable lesser systemic indignities when they seek medical attention, such as waiting room chairs and blood pressure cuffs that are not big enough. But the most severely obese patients raise another important question: Where do we draw the line? Should all aspects of our healthcare system be designed with thousand-pound human bodies in mind? Or is there a limit beyond which people cannot expect to be saved from their own body fat? Ultimately, these are systemic and ethical problems that are not easily solved.

The night before I left Barrow, I had blubber for dinner. Herman had given me some leftover muktuk from his ice cellar, and I stood in the small kitchen in my lodging, cutting off the black skin and slicing the pale blubber into small chunks. I planned to render the pieces of fat and use the resulting oil to fry sliced potatoes—a dish that could be called "whale fat fries."

Fat is the human body part most heavily laden with emotion and judgment and most colored by culture, but as I watched the chunks of muktuk sizzling in a frying pan on the stove, I thought about how, in the Arctic, fat is not the enemy but the hero. The Iñupiat and their ancestors came to inhabit the seemingly uninhabitable because of it. Though most Iñupiat people no longer depend on it as they once did, new research is proving blubber to be far healthier than most things for sale in Barrow's supermarket. The Centers for Disease Control and Prevention (CDC) is now encouraging Alaskan Natives to increase their intake of traditional foods, despite their fat content—modern science is finally catching up to ancient common sense.

Every individual human body is anatomically constructed in part by the socioeconomic and historical context in which it lives. There is growing awareness among medical professionals about our own fat biases, as well as systemic problems in our industry, which is the first step toward improvement. After decades of biased nutrition science and false starts, the medical establishment may finally be laying off the fat-shaming—both on the body and in the diet—and finally taking into account the complexities of metabolism, genetics, and food availability in the way our bodies are shaped.

I wondered what traditional Iñupiaq mythology would have had to say about blubber's sacredness. Unfortunately, all information about the Iñupiat's traditional religion has been lost, with nothing left of an ancient outlook that was shaped by the land and sea as much as the Iñupiaq diet and genetics were. Still, as I ate my dinner, I imagined Herman's ancestors singing songs about the magic of dazzling and gigantic sea mammals, the glory of summer's explosion of life, and the exuberance of piles and piles of healthy, delicious, life-sustaining fat.

11

LUNGS

In 1969, the US Department of Agriculture (USDA) began a study of livestock lungs to determine once and for all if they were fit to be called human food. While all meat and animal organs bound for human consumption undergo some type of inspection under USDA regulation, this particular study embarked on a much more thorough examination than usual. Government pathologists collected several hundred cattle lungs from various slaughterhouses and dissected them thoroughly, looking through the entire air passages branching into both lungs, all the way down to the deepest microscopic air sacs, the alveoli. Their goal was to scientifically determine whether lungs were safe to be sold and consumed by people.

In a large majority of the samples, the pathologists discovered dust and mold spores that the animals had inhaled, along with small amounts of contents from the animals' stomachs that ended up in the lungs, aspirated either before or after slaughter. Even more alarming to these government scientists was the fact that many of the contaminants were discovered in the smaller airways deep down in the lungs, where a meat inspector might never see them during a routine examination. They repeated the study with sheep and calf lungs, and they found the same impurities.

To USDA bureaucrats who had ordered the study done, there was only one solution: in order to balance consumer protection with the impracticality of dissecting every animal lung, they decided that the sale of lungs as food should be banned altogether. Though people had been eating lungs for eons, in 1971 the rule was inscribed into the US *Code of Federal Regulations* (C.F.R.), and it remains on the books to this day. Dogs can still eat lungs—many dog owners swear by the chewy lung cubes frequently sold at pet-food stores. But for humans, lungs are on the USDA's very short and lonely culinary black list, as are lactating udders. Non-lactating udders, on the other hand, are still permitted.

As I read about this history in USDA documents and the federal code, I was confused by the law's medical reasoning. Each of us breathes in fungal spores and dust all day, every day, and while it might be unhealthy to inhale large amounts, eating them did not seem particularly dangerous to me. And the same went for stomach contents—people still widely eat tripe, the lining of an animal's stomach, which is perpetually bathed in the same "contamination" that worried the pathologists who found small amounts of it in the lungs. To me, the regulation against consuming lungs seemed to have more to do with legislating what is "gross" than it did with protecting people from what is actually unsafe.

Though hidden within the chest, lungs are connected to the world outside the body, which makes them unique among internal organs. The lungs are like the skin—a boundary with the rest of the world—but turned outside in. The air inside our lungs is connected to and continuous with the atmosphere that moves in and out with breathing. Each person's lungs enclose a tiny bit of personal atmospheric real estate, and it is through this air that our bodies extract needed oxygen and rid themselves of spent carbon dioxide.

As government pathologists discovered in 1969, this fact means that lungs are exposed to all of the air's dust and mold spores. The only pure lungs are those of unborn babies who have yet to breathe, but the moment a newborn takes its very first breath, those perfectly unblemished, virgin lungs are muddied by the atmosphere's adulterants. Every subsequent breath only multiplies the accumulated grime. Lungs generally handle the daily load quite well and constantly clean themselves of contaminants; otherwise, living in the fungal blizzard of the earth's lower atmosphere would be impossible.

That lungs are a bridge between the body's inside and outside was highlighted for me the first time I saw a pair of human lungs in anatomy lab. To get to them, I had used a scalpel to slice through the skin and fat until the sharp blade hit hard bone, and then a circular saw to grind away at ribs on either side of the chest. Bone dust showered the organs in an already-open abdomen. After a few more cuts, my fellow medical students and I removed a section of chest wall to reveal the central heart framed by two lungs. My cadaver's story became clearer immediately—his lungs were speckled a dark gray like two fluffy thunderhead clouds.

Healthy lungs are colored a juicy pinkish-beige, but decades of soot and smoke from burning tobacco had blown through his airways and painted his lungs the color of ash. We were never told any particulars of either medical or personal history about our cadavers—the only clues to their lives were discovered by opening them up, and we were free to imagine the details. It took only a quick glance at his lungs to fire my imagination about the daily life of a mystery man I met only after he had smoked his final cigarette. I envisioned a steady stream of cigarettes nuzzling his lips over the years, and thought about what his lungs would have sounded like if he had been my patient. I probably would have heard the crackling and wheezing sounds of chronic inflammation common in smokers.

My cadaver's lungs were only a more extreme example of what USDA pathologists found so concerning in animal lungs. Humans inhale far more pollutants than animals do, partly because many of us purposely breathe in burning vegetal matter such as tobacco and marijuana. Lungs stand at the front door of our bodies and bear the brunt of air pollution, our reliance on burning wood and fossil fuels, and our exposure to industrial by-products like asbestos, silica, and coal dust. Archeologists know when our human ancestors began cooking food over fires by the presence of black soot in their lungs, another postmortem story told by our bodies long after we've abandoned them. Cigarette smoking is the same daily pollution of an internal organ, brought to its most excessive limit imaginable.

The color of my cadaver's lungs jibed with another discovery we made in his heart—his coronary arteries were calcified, and they crunched like hard caramel between my gloved fingers. Cigarette smoking is among the strongest risk factors for coronary artery disease, and he very likely suffered from the same. Toxins coming into the lungs often enter the bloodstream and affect distant organs, and the residues of tobacco smoke in particular injure blood vessels throughout the body, including the coronaries. Perhaps he had lived with intermittent chest pain as a result, or had the scare of a heart attack once, or maybe repeatedly.

Looking inside dead bodies, I realized, meant reading the footprints of past disease and gaining a view into the life that a person lived before ending up on a cold gurney to be picked at by nervous medical students. Each body, and every organ within it, has some story to tell, one that weaves a tale of sickness and health from one organ into another.

My cadaver was not a healthy man before his death, and he likely suffered from a host of chronic diseases underlying the acute illness that eventually killed him. Such pathological findings are the same type that would be of interest to USDA inspectors conducting the usual postmortem examination of animals—they look for evidence of disease. Eating body parts of ill animals can pose a real medical risk to humans, but in the 1969 study of animal lungs, pathologists did not find that most of the organs were sickly. If they had found evidence of disease, it would have justified condemning those specific, individual lung samples. But all they found was evidence that animals breathe.

Early on in anatomy lab, as I was first getting elbow deep in my cadaver's abdominal fat and neck deep in the Latin names of body parts, I decided to visit a slaughterhouse. I wanted to learn more about how cuts of beef compare to human muscles. I found a kosher slaughterhouse in central

New Jersey, deep in the state's industrial heart, and I called up the owner. After expressing surprise at my request and asking a few questions to convince himself that I was not "some crazy vegan or something," he agreed to let me visit on the next slaughtering day. And while I had muscles on the mind, the theme of my visit would turn out to be all about lungs.

On a crisp autumn morning, I drove along the New Jersey Turnpike past oil refineries, gas stations, and tractor-trailers to the slaughterhouse. When I opened the heavy metal door to the building, I could hear the rattling of chains, the booming sounds of chain saws, and a chorus of cattle mooing. The scent of barnyard hung in the cold air as I walked through the front office toward the dreadful sounds coming from beyond.

The slaughtering had already begun. I saw rabbis with long gray beards and thigh-high rubber boots standing around a large wooden table, examining mounds of shiny flesh. Workers, primarily Black and Hispanic, wielded huge motorized butchering saws and moved hanging quarter-cows along tracks in the ceiling. Each steer was led into the building from the outside lot through a narrow chute leading directly onto the slaughtering platform. Chains were then fastened to its back legs and used to slowly lift the animal off the ground. Just as the front hooves left the concrete floor, a long, final *moo* built in volume and echoed off the grimy industrial walls. With one swift slice of the rabbi's knife to the animal's neck, a slick of blood hit the floor with a loud splash, and the animal was dead before the echoes of that last *moo* had finally faded.

I walked among the hanging slabs of beef and saw quarter-cadavers, recognizing the same orthopedics of muscle and bone that I had seen in anatomy lab. Underneath our skins, humans and cattle are both glistening red outlined in white, strung like puppets by the names of a dead language.

The rabbi actually doing the slaughtering seemed less busy than the others—in between animals, he mostly stood around cleaning the blood off his long knife. His beard was neatly cropped, and his yarmulke held tightly to his short brown hair. I asked him about what the other rabbis were doing.

He explained that Jewish traditional dietary law, or kashrut, provides a guide to the proper dissection of meat and diagnosis of its cleanliness. I knew the basic rules of kashrut: keep dairy and meat separate, and avoid shellfish and pork. But there is another criterion that is less well known, he explained: severe pneumonia during an animal's life can make an animal no longer kosher.

In healthy animals and humans, as the lungs expand and contract with each breath, they slide freely against the pleura, a layer of membrane surrounding the lungs and lining the inner side of the chest wall. But when the two surfaces are inflamed by a bad bout of pneumonia, they stick together like an unlubricated piston in its shaft. As the pneumonia heals, a scar forms at the spot where the lung got stuck—a band of white fibrous tissue attaching the two surfaces. The shochets—those trained in kashrut's version of a USDA inspection—were carefully examining the animals' lungs and looking for these telltale signs of pneumonia. Called adhesions, these scars were the footprint of past disease, and each was a potential degradation of kashrut. According to Ashkenazi Jewish tradition, the number and size of these adhesions determine the grade of kosher, with the highest level called glatt, meaning "smooth," a description of the surface of an animal's lungs that are free of the roughened scars.

Most important, the shochets must determine whether there is a hole hidden within a scar that reaches straight through the lung. As a carcass hung freshly killed and cut open, a shochet slid the lungs out of the chest cavity. He walked back over to the examining table, his hand grasping the trachea as two fleshy lungs dangled below. He placed an air hose into the animal's trachea and inflated the lungs with a rush of air. They doubled in size like two large loaves of bread rising abruptly. The shochet then cupped his hands around one of the scar tufts on the lung and filled his hands with water, being careful not to let any drain out. If there was a hole within the scar, air from inside the lungs would bubble up through the water, as when a mechanic investigates a flat tire for the puncture site. Such a hole from the outside into the body's inside proves the animal is not intact and therefore its entire body is not kosher, with bubbles as the definitive diagnostic criteria.

Kashrut's concept of cleanliness and health seemed to rely on the sanctity of a barrier between the inside of the body and the outside world. Maintaining cleanliness means keeping the outside out, much as people in many cultures remove their shoes before entering a house or a place of worship. When animals or humans breathe in air and atmospheric schmutz, they enter our lungs and whoosh all the way down to the alveoli—but this is not truly *inside* the body. The air in the lungs is still continuous with the external atmosphere. The real threshold of the physical self is the lining of those deep alveoli, and a hole connecting the inside of the lungs to the pleura is a way for the dirt of the outside world to get in, truly *inside*, the body, and once that sacred barrier has been breached, innocence and purity are soiled.

For the kosher postmortem inspection of an animal, the lungs have a unique primacy—they hold the singular key to the purity of every part of an animal's body, even its rump roast. In the past, shochets examined eighteen different body parts to make a determination of kashrut, looking for defects of all kinds, but experience over centuries showed that the lungs offered by far the most bang for the buck. A large enough proportion of all defects found were in the lungs, obviating the practicality of examining the other seventeen body parts, except in special circumstances.

It made anatomical sense: as the organ standing guard at the body's entrance and suffering the microbial blows of an outside world teeming with infection, the lungs serve as a proxy. The kosher version of dissection exalts the lungs above all other organs, and when they show signs of disease, the animal's entire body is considered unfit for human consumption.

The moment that animal lungs in the United States are removed from the body in which they breathed, they enter a uniquely restrictive legal context. I searched government records at the USDA, the National Archives, and the National Agricultural Library for any historical context that would explain what prompted the initial USDA investigation, but I found nothing. Several people at the USDA could also find no record for the original impetus that led to the ban.

I reached out to legal experts, most of whom had never heard of the law. According to Theodore Ruger, dean of the law school at University of Pennsylvania and an expert on food regulation, the law against lungs is a perfect example of regulatory inertia. It has remained on the books simply because no significant push exists to overturn it. As Ruger explained in an email, "The general phenomenon of outdated regulations is commonplace, particularly in policy areas where the impact is relatively minor so there is not enough countervailing lobbying energy to fix the law." With little hankering from the public, and not much of a push from industry, there is little impetus for the USDA to spend resources on reassessing the safety of eating lungs.

The one significant push that exists to overturn the lung ban comes, surprisingly, from Scotland. Lungs are a classic ingredient in haggis, a beloved food consisting of a mix of internal organs—typically heart, liver, and lungs—stuffed into an animal's stomach and cooked. Haggis enthusiasts insist that lungs are an irreplaceable ingredient in the traditional dish, providing the necessary crumbly texture.

I tried haggis during a trip to Scotland—it was served as part of a massive hotel breakfast along with various other meats, cheeses, and bread. The haggis came as two circular slices, light brown and crumbly, speckled with embedded oats and barley. When I put a small piece into my mouth, the only thing that I could taste was liver, and at the time, I had no idea that I was eating an organ prohibited in my homeland. The liver's overpowering taste drowned out the taste of nearly anything else, and lungs, I learned later, are like the body's tofu: they acquire the taste of whatever else they are cooked in.

The United Kingdom has been pushing to overturn the US ban on the sale of lungs since it was first established in 1971, because it prevents the importation of authentic haggis from Scotland. According to Owen Paterson, the UK's former secretary of state for Environment, Food, and Rural Affairs, the ban is not grounded in legitimate health concerns. He pointed to the fact that the United Kingdom exports haggis to many other countries, and that, to his knowledge, there has not once been a complaint regarding the safety of the product. He groused, "The US gets annoyed with the EU whenever it blocks US products on spurious grounds. So, the US should not do the same then, as they do with lungs." He sounded exasperated, and I couldn't blame him.

Over the course of several trips to the United States from 2012 to 2014, Paterson took up the decades-old push to overturn the law against lungs but never succeeded. He admitted that haggis exports to the States would likely never approach the trade in Scotch whisky, which is the biggest food or drink export of the UK "by a mile." Still, he expressed hope that in the post-Brexit context, a bilateral trade deal would be inked between the UK and the US, and in those negotiations, haggis would be a primary point of discussion.

But it's not just the Scottish that eat lungs. Prior to the ban, lungs were commonly eaten in America by immigrants from various parts of the world. Lungen stew, for example, was a lung soup eaten by Jews from Eastern Europe, though nobody in my Jewish family had ever tasted it. One family friend in Brooklyn with whom I spoke was very proud of her lungen stew recipe, though she hadn't made it in forty years. She recalled how she cut away airways and blood vessels, parboiled the lungs, and then removed their "membrane" (pleura). She chopped them up with onions, carrots, and sometimes potatoes, and the result was, she boasted, "terrifically delicious."

My childhood friend's mother, a daughter of immigrants from Tuscany, Italy, shared with me a recipe for *coratella*, a traditional dish of offal including lungs, which is sold from carts in her father's home town of Lucca. The recipe required beating the air out of lungs before cooking them, a reminder that lungs are, by volume, more air than flesh.

Many people I spoke to could recall lung dishes being sold at diners and restaurants in decades past. Another friend's father recounted always ordering lungs as a child when his family ate dinner at Greenspan's, a now-closed restaurant in Bayonne, New Jersey. He remembered it was cooked in a thick, gravy-like sauce with tender chunks of meat that were visibly identifiable as lung tissue. The dish was rich, unctuous, and dense, the lung "spongy and just delicious." He had no idea that lungs were illegal until recently when he, in an attempt to re-create the dish from his childhood, asked his butcher for lungs and learned of the ban. Like the others, he lamented the disappearance of lungs, and all of them reminisced fondly about a beloved food from the past.

After months researching the topic and discussing delicious-sounding lung dishes, I resolved to try lungs again (knowingly this time). If there was a black market for animal lungs in the Northeast, I could not find it, though not for lack of trying. I would have to wait until the next time I left the country. While preparing for a trip to Israel, I heard from a friend who worked as a tour guide there that a particular Bulgarian restaurant had lungs on the menu.

After a long airplane trip and a few days of visiting my in-laws, I found the restaurant called Monka in Jaffa. The restaurant was empty and the waitstaff sat outside smoking cigarettes as we entered, a cryptic reminder of the lungs from my days in anatomy lab. I sat down with my wife, Anna, and my one-year-old son at a plastic table. At the bottom of the menu's entrée list was a dish called "lungs in juice." I had finally found the forbidden fruit of internal organs.

The waiter came over and took our order. My wife went first and ordered a salad—the waiter scribbled on his notepad and nodded. When I placed my order, he stopped. With his pen's tip motionless and still pressed against his notepad, he raised his eyes and looked at me: "You know what it is, right?"

With my wife translating, I answered, "Yes, and I'm very excited because it is not sold in the US."

He explained that the lung dish is almost exclusively ordered by elderly Bulgarian, Romanian, and Turkish customers, and never tourists like us— hence his surprise. Since Monka opened in 1948, the same year of Israel's founding, the restaurant has been known for its lung dish, as well as an intestine soup that, these days, is far more popular than the lungs in juice.

When I asked why lungs aren't more popular, he answered, "Because people do not know how to cook it." Curiously, he also didn't know the recipe—his brother was Monka's resident lung chef. The only thing he knew was that the organ had to be simmered in tomato juice for a *very* long time.

When the dish arrived, one half of the plate was covered with a mound of rice splattered with beige, saucy beans. The other half of the plate contained chunks of lamb lungs. The shiny flesh was a light pinkish brown with no contamination or gray spots of tobacco soot discernible to my eye. The pieces had a sheen on them, the pleura, which reflected the restaurant's overhead lights, giving it an iridescent glow.

As I cut into the meat, for a moment I was back in anatomy lab. I dissected with my knife and fork and discovered branching airways coursing through the soft flesh connecting all of the lung's empty pockets. These spaces would have been filled with atmospheric air in life, but in death, they were filled with a thin tomato gravy and draped with bits of translucent cabbage. The lungs cleaved easily into cubic sections about half an inch in size that had once fit together neatly to constitute the whole organ, but the lung lobules had broken down and fallen apart with prolonged cooking.

As I shoveled bits of lung into my mouth, along with rice and silky soft beans, I thought about the fact that I was also probably eating mold spores and dust that the lamb had inhaled during its last few days on earth. But those "adulterants," as the USDA federal code labeled them, would cause me and my family no harm—they might even add to the meal's nutritional content. Besides, much of the debris that people inhale, including spores and dust, ends up being swallowed anyway—the same mucus elevator that helps lungs clear out aspirated food also continuously flushes inhaled particles upward toward the throat. We unconsciously swallow this mucus and the refuse it gathered from the lungs all day, every day, in a unique instance of the human body's programmed auto-cannibalism. In other words, we all constantly ingest the same contaminants from which the regulation against lungs supposedly protects us. Furthermore, given how abundant spores and dust are in the air, I would guess that every bit of food that any human has ever put into their mouth probably contained plenty of mold spores and dust.

The lungs were tender and delicious, their taste a savory mix of meat and tomatoes. As my son had not developed any food preferences yet and had no sense of revulsion, I took advantage of his youthful ignorance to share my meal with him. He loved it, and so did I. Even Anna found it not nearly so objectionable as some of the other organ-based foods I had offered her in the past.

During life, all of an animal's internal organs exist alongside and bound to one another, but the moment that they exit a slaughtered animal's body, they each become subjected to the nuances of socioeconomics, culture, and what any given group of people define as cuisine. In many cultures, lungs are seen as food for people of the lower classes, and food science has a history of reflecting popular preferences thinly disguised by a facade of objectivity. Choosing which organs any of us eats has more to do with culture and class than it does with anatomy and physiology, and more to do with habit and tradition than taste.

When bodies are opened after life, the lungs of every human and air-breathing animal reflect the relationship between the body's insides and outsides. Whether in a dissection lab or a kosher abattoir, those lungs will tell a story of a lifetime of air moving in and out, a tale of exchange and consequences. Laws, like lungs, can be examined closely, too, and perceptions of them can change as well. In my opinion, lungs are perfectly healthy to eat if you want to—you'll just find it very difficult to do so (legally) in the United States.

12

EYES

The last thing Fred said he saw before it happened was a shower of sparks. A custom metalworker in rural Pennsylvania, he had been holding a grinder's coarse spinning wheel against a new set of door hinges he was fashioning, while a salvo of white-hot metal fragments arced across his workshop and scattered on the floor like spent shrapnel. As he shifted the grinder's position to attack the hinges from a different angle, the wheel touched down again and he instantly felt an agonizing sting in his right eye. Despite the pain, he drove from his home to the ER where I was working with his right eye shut.

I found Fred pacing back and forth in his bay in the ER, furiously re-counting what had happened.

"It feels like that piece of metal is scratching the shit out of my eye!" he shouted.

His short, dirty blond hair was a mess of sweat, and he smooshed the palm of his grease-covered hand into the hollow of his injured eye. Fred stood a head taller than me and made me nervous as he plodded around the room like a wild animal wounded in its most sensitive body part. Despite my pleading, he refused to sit down.

Diagnosing and treating eye complaints in patients clashes with a simple fact about the human body—every fiber of our being is dedicated to keeping things out of our eyes, including the hands of examining physicians. But I worried that the pressure of Fred's right hand, along with the grease on it, might further damage the eye that had already been assaulted by flying metal shards—people too often reinjure their eyes by rubbing after an initial injury or irritation. With a promise of pain relief, I finally coaxed him to sit down on the bed and managed to gingerly pry his hand away from his face. He could barely keep his right eye open, but for a brief moment I glimpsed the damage: a bloodred color glimmered between his spasming eyelids.

All ER doctors know that injuries to the face are exceedingly common. A colleague specializing in ear, nose, and throat surgery once told me that alcohol-related facial injuries put his children through college. When I began working as an ER doctor, I, too, found that my patients' faces were the body parts that sustained the brunt of humanity's violent and drunken stumbling tendencies. But after seeing many patients with facial injuries, I noticed one consistent feature: direct hits to the eyeball are surprisingly rare.

A patient's face could be bruised beet purple, their eyelids massively swollen to the point that they cannot even open them, their nasal bridge flattened and the sinus beside it smashed and filled with blood and bone fragments—but once I manage to pry open their eyelids, I typically find the eyeball sitting comfortably and perfectly uninjured in its cushion of eye socket fat. When I see that it swivels freely in all directions, that its central black pupil shrinks in answer to my light, and that its surface is white and quiet with no hint of the angry red color I had glimpsed in Fred's, I know the eyeball has escaped injury once again.

The human face is expertly crafted to protect our all-important and oh-so-delicate eyeballs from injury. Each eyeball is set deeply into a bony crater known as an orbit, which is outlined by a circular ridge of bone that protrudes like a shield on all sides. Above the eye is the brow line, a bony crest topped by an eyebrow and jutting outward like a protective visor, and beneath the eye is the cheekbone, a matching prominence. The nasal bridge defends eyeballs in proportion to its widely varying protuberance, and the eye's outer edge closest to the temple also rests in the shadow of a solid and safeguarding bony corner of the face. The terrain surrounding our orbits is clearly well designed to fend off projectiles, fists, and table corners, and it usually succeeds.

Our bodies are also programmed with well-honed reflexes to protect our eyes. Reactions like flinching, or blocking incoming objects with our hands, are so deeply ingrained that they are difficult to temper. Our eyelids involuntarily blink just in time to keep out flying debris, and they resist being held open even when the brain knows that an object in close proximity is safe. These instincts made it difficult for me to learn how to put contact lenses into my own eyes, and I'll admit that putting in eye drops is still a challenge. It's especially hard for patients to overcome their knee-jerk reactions when I have to go searching in their eyes for minuscule objects that sneaked in.

If the body's facial architecture tells us one thing, it's that the eyes are supremely important and just as fragile. Yet despite all these precautions, things still sometimes slip past the goalie and get into our eyes.

And when they do, such "foreign bodies" are among the body's most bothersome phenomena. When caught in the eye, even a barely visible speck of dust that would be hardly noticeable anywhere else on the body's surface becomes disproportionately annoying and utterly impossible to ignore. Even the measliest eyelash—a specialized hair attached to the edges of our eyelids to keep out debris—when shed and lodged in the eye becomes an urgent distraction.

The perpetually moistened surface of the eyeball is one of the body's most sensitive patches of surface real estate. In this way, eyes are similar to testicles—trauma to those organs causes pain that is far out of proportion to the force applied or the damage done. Ovaries, another paired organ, are sensitive, too, but they are protected within the pelvis' bony ring from most direct hits. The human body's daintiest and most fragile parts seem to always come as a pair of delicate orbs.

When patients have something "in" their eye, it is usually not actually *within* the eyeball's spherical globe. Foreign bodies are instead often somewhere *on* the eyeball's surface, underneath the eyelid or within one of its creases. And that is where I went looking for Fred's speck of metal.

I grabbed a small bottle of numbing solution from the "eye box," a collection of instruments and diagnostics that ERs always have ready to use. Following my instruction, Fred reclined on the hospital bed, tilted his head back, and looked up at the ceiling. I dribbled a few drops into the inner corner of his right eye, where it pooled momentarily before tumbling between his eyelids and sprawling across the surface of his eyeball into a watery glaze. The pharmacologic ability to completely numb the surface of a patient's eye is one of the most useful tools in all of medicine. Without it, removing a foreign body would be far more brutal for both patient and doctor.

After a minute the drops kicked in and Fred could comfortably keep his eye open. With the searing pain relieved, his tense facial muscles also finally relaxed. I grabbed a light from its resting holster on the wall behind him and bent over for a look. As I reached toward Fred's right eye with my hand, his head jerked away, and his restless torso told me he was still having trouble letting his instinctive guard down.

"Try to hold still," I said, my face hovering inches away from his. He smelled like motor oil.

I spread his eyelids with the fingers of my left hand and scanned the eye's surface with my light, looking for a small dark dot that appeared out of place. The black pupil in the middle of his eye was framed by a brown ring of iris, and both were covered by the cornea, a transparent dome that reflected my light. I saw no dots or scratches anywhere. I looked closely at the white of Fred's eye, patches of it stained the dark pink of a dragon fruit rind—like any other injured body part, a traumatized eyeball experiences a rush of blood to help with healing. His eye was also overflowing with tears, his body's attempt to flush out the intruder.

In any thorough search for foreign bodies in the eye, a patient's eyelids must be flipped over—I've discovered many objects that have come to rest on their underside, affixed and scratching the cornea again and again with every blink. I removed a sterile Q-tip from the eye box and pressed its soft, bulbous end gently onto the front of Fred's upper eyelid. With my other hand I grabbed the lid's eyelashes and pulled upward, folding the lid back on itself to reveal its moist, pink underbelly. This portion of the eye exam was difficult for me when I first learned it—I found tear-soaked eyelids slippery and difficult to grab, and I hesitated to forcefully tug and fold them. The sensitivity of my own eyes made me reluctant to manhandle another person's.

Over years of practice, my technique became faster and more precise, as it did with many other uncomfortable medical procedures. I learned to be appropriately brutal with eyes, though I would never approach the brutality of ophthalmologists, who perform surgery on eyeballs, grabbing them with tweezers and cutting right through them.

I thoroughly searched every crevice in Fred's eye, but I found no dots or shards of metal. I shut off my light and stood up.

"I don't see it," I told Fred. He glared at me with a suspicious, disbelieving look that I recognized from many other patients who were convinced that something—often a contact lens—was still stuck in their eye. Some patients have even expressed concern that a lost contact lens might have slipped all the way behind their eyeball. In these situations, I perform an exhaustive and typically fruitless search, and then suggest to them that the contact lens likely already fell out and they are just feeling the leftover irritation. I also explain that the eyelid crevices are dead ends and objects cannot get past them. But they often don't trust me, and their disbelief is written on their faces—just as it was on Fred's.

The sensation of having something in the eye is so visceral that a person's immediate instinct is to trust that feeling—and to distrust the doctor, who might have simply not looked hard enough for it. For many patients seeking help for something stuck in their eye—and the same goes for patients who feel something stuck in their throat—there is usually nothing there at all except a scratch.

But Fred's case was different. I told him that there was one other possibility—the sliver of metal might actually be inside the globe of his eyeball. Grinding wheels spin rapidly, and the white-hot metal shards they throw off travel at high enough velocities to penetrate the eyeball. Such objects, known as intraocular foreign bodies, are uncommon, and when they do occur, there are often telltale signs. A punctured eyeball can rupture like a popped water balloon and deflate as the viscous goo inside leaks out, causing the eyeball's white surface to fold and buckle. The pupil can also be tugged out of its typical circular geometry into the shape of a teardrop or a miniature black river running into the surrounding colored iris.

But a violated eye can also look normal, as Fred's did, his globe appropriately plump and turgid and his pupil perfectly round. But I knew that what I could see with my own eyes was unreliable, and many patients with intraocular objects leave the ER with a diagnosis of only a scratch on the cornea. A missed intraocular foreign body can devastate a person's vision if left inside. The only way I could know for sure was to peer into Fred's eyeball with a CT scan. I placed the order into my computer and awaited the results.

Though a physician's job appears, on its surface, purely scientific and technical, interacting with patients is still an essentially social act. Some of the most important lessons in my medical training came not during class or while looking at microscopic slices of the body but by watching artless doctors interact with patients and their families. Often it was while participating in end-of-life discussions, when the attending physician's mannerisms or poor communication skills clearly flustered a patient and their family; or the time I saw an attending in a rush use his stethoscope to listen to the heart and lungs of a patient who was sitting on the toilet, clearly embarrassing and dehumanizing the poor woman. These are the most egregious examples I have seen in my career, but they stayed with me and helped shape my own style of practice.

With one particularly awkward doctor, I myself was the patient. It was a standard checkup during medical school, and the doctor stared at his computer screen the whole time, his fingers hunting and pecking for letters on his keyboard throughout our conversation. I was aware of the growing burden of electronic documentation on physicians like him, and I wondered if he struggled to get all his clinic notes finished each day—his painfully slow typing supported this hypothesis. In a busy and over-technologized world, the eyes of physicians, like everyone else's, are glued to screens. It was an exasperating experience, and it taught me that to be seen is to feel human.

The practice of medicine requires more than just the ability to make yes-or-no diagnoses—physicians must also meet patients on a human level and build trust. Eyes become crucial here since they carry a heavy social weight in the doctor-patient relationship, just as they do outside of healthcare. The eyes of other people are a natural focal point for our own gaze—when we look at others, we generally look at their eyes, as if we believe each individual self rests, more than anywhere else in the body, within the organs of vision. Eye contact is the epitome of human connection, and one of the first social behaviors seen in young infants. As a result, eye contact plays an outsized role in bedside manner, and too much of it can be as awkward as too little.

Doctors are often guilty of focusing on a patient's internal organs and lab data while altogether missing the very basics of human-to-human interaction. Whatever malfunctioning body part brings a patient to my medical attention, I learned to first interact with their eyes when introducing myself and chatting about their condition. I trained myself to be aware of how much eye contact I make with a patient and their loved ones, and I aim for just the right amount. I eventually make skin-to-skin contact with all patients, but the touchless contact of eyeballs is, in many ways, more intimate—even from across a room.

Eyes are not just sensory organs but also ones that actively communicate complex emotions. Direct unwavering eye contact can be a threat or a lustful invitation depending on the context. The anatomical terrain surrounding our eyeballs adds to the message—the minutely muscled and highly dexterous eyebrows and eyelids can rise together in surprise or sag with disappointment, while skin over the nasal bridge can wrinkle in anger. This communicative rippling of the human body's surface gives a physician important information about how a patient is feeling. For instance, the shape of skin around Fred's eyes as I left his room told me that he thought I was a complete moron.

Though humans can express to one another an infinite array of thoughts and emotions using the voice box, eyes are somehow more intimately intertwined with a person's innermost essence. I once heard an interview with a woman who had lost her child at a young age—she said she could detect a similar loss in the eyes of other people. Like a second voice coming from our faces, eyes are often the more honest organ.

An hour passed before Fred got his CT scan completed. While waiting for the radiologist's official interpretation, I pulled up the image on my computer screen. Fred's eyeballs appeared as black circles on the scan, and sitting in the very center of his right eye was a bright white dot. I had finally found the tiny sliver of metal. And I realized I should not have been as thorough and brutal while examining a punctured eyeball.

I told Fred the news, and could instantly see worry in and around his eyes. I explained that because it was inside his eye—actually *inside* the eyeball—I could not remove it. He needed an ophthalmologist and I would need to transfer him to another hospital. Practicing medicine requires more than just keen observation skills—it involves knowing when I cannot trust my own eyes and should turn to the penetrating vision of a CT scan or other kinds of medical imaging.

An ancient poet once said that the eyes are a window to the soul, and I have found this to be true in the practice of medicine—and in some of the most critically ill patients, eyes can also be windows to a lifesaving diagnosis.

One morning while working in the ER in rural Pennsylvania, a frantic phone call came in. The caller was speeding toward the hospital in a car because his friend in the passenger seat had turned blue and stopped breathing. A few nurses, a tech, and I prepared for a "car retrieval" and headed toward the curb in front of the hospital. When the car screeched to a halt in front of us, we wheeled our stretcher toward the passenger door and dragged from it a large man, his face as blue as the sky and his body as limp as a newborn baby that had not yet taken its first breath. I grabbed a fistful of the sports jersey he wore and helped pull him onto the stretcher, the sunglasses on his forehead knocked askew as he flopped lifelessly. We wheeled him quickly into the ER, where a nurse placed an oxygen mask over his mouth and nose, and another inserted an IV into his arm.

Given the man's serious condition, there was an urgent need to figure out the cause of his problem. I did not reach for a stethoscope, a blood pressure cuff, or a reflex hammer—instead, the first thing I grabbed was the same light I'd used to look into Fred's eye. For the most precarious patients, looking into their eyes quickly gives me a few crucial pieces of information.

I spread open the man's eyelids one at a time with the fingers of my left hand and shined the light at his pupils. I was first looking to see if he had a "blown" pupil—one that is extremely large and refuses to shrink in response to light. This ominous sign announces dangerously elevated pressures inside the skull, possibly caused by a massive brain bleed. When I find a blown pupil in a comatose patient, I suspect that something catastrophic has happened inside their skull, and I immediately call for a stat head CT and a neurosurgeon.

But neither of the man's pupils was blown. In fact, I found the opposite—both his pupils were as small as pinpoints, and they told me instantly what had happened: an opioid overdose.

Opioids like heroin and fentanyl suppress consciousness and slow breathing, sometimes bringing them to a complete and fatal halt. But these substances also act on the eyes, causing pupils to shrink and look like two black poppy seeds—the same seeds of the opium plant that originally bestowed on humanity a family of drugs with unrivaled power in pain-killing, as well as in causing addiction and dependence.

As soon as I saw the man's pupils, I asked the nurse to draw up a dose of naloxone, the antidote to opioids. One squirt went into his nose, and the nurse injected another few milligrams intravenously. Within thirty seconds, his head lifted up off the bed, his eyes opened wide, and his face turned from blue to pink, like a newborn beginning to breathe. He looked around at the crowd of people dressed in hospital scrubs surrounding him, a bewildered look in and around his eyes. Naloxone's instant effect confirmed the diagnosis. His once-drooping eyelids suddenly stood bolt upright, and his pupils widened into a pair of black lentils.

"You're in the ER," I told him, attempting to answer the unspoken questions I saw written on his face. "Did you take anything today?"

"I had some heroin," he said. "But I took my usual amount, and I've never OD'd before." He wondered aloud if perhaps the heroin had been cut with fentanyl, a much more powerful and deadlier opioid.

He wasn't the first opioid overdose I saw that week—I cared for patients like him often, working in that part of Pennsylvania. The region has been hit hard by an epidemic that has ravaged the country, and its victims frequently end up in the ER in similarly dire condition. When evaluating such lifeless patients, seeing pinpoint pupils in their eyes makes me relax—I instantly know the problem, and its solution is a simple one. There will be no need for a breathing tube, an emergent CT of the brain, a page to the neurosurgeon on call, or a transfer to another hospital. Instead, a quick and simple squirt of naloxone, and the coma is reversed in one of modern medicine's most magical and instant resurrections. The eyes may be a window to the soul of a human being, but they are also a window to the biochemistry of a brain distorted by mind-altering, pupil-reshaping substances. In this way, they are also organs of communication, especially for the comatose and speechless.

I encounter opioid overdoses less frequently now, ever since the antidote became more widely available. First responders and police officers carry it, and Pennsylvania and other states recently began allowing pharmacists to dispense naloxone without a doctor's prescription. As a result, the overdosed are more commonly revived out in the community than they are in the ER, often by their own family members, who are more aware of when and how to use it. Once revived, those who have overdosed often become alert enough to refuse further medical care or transport to the hospital, and I never see their telling eyes with my own.

A second but lesser-known drug epidemic is that of methamphetamine, and it, too, shows up in my patients' eyes. Meth, as it is commonly known, causes the opposite effect to that of opioids—instead of becoming sedated and sleepy, individuals become agitated and energized, and their pupils widen. One patient was brought into my ER after a fall from a second-story window—the paramedics had found his body crumpled and folded into an unnatural shape at the bottom of a narrow alleyway. The patient told me that an argument with his girlfriend had possessed him to throw himself out the window for reasons he could not quite explain. When I asked where it hurt, he claimed he had no pain at all—which made little sense given the height from which he had fallen and the near certainty that he had broken something.

But when I looked into his eyes, I found a clue—both his pupils were gigantic, each a yawning black hole surrounded by only the thinnest rim of light blue iris. They looked the polar opposite of ones shrunken by opioids. Meth is a powerful stimulant, and it revs up the body like a shot of adrenaline—blood pressure and heart rate skyrocket, and the pupils are activated into their vigilant state of dilation, a wide-open window into the patient's story.

I asked him if he had smoked, snorted, or shot any meth that day. He denied it, but I still suspected otherwise, and I knew that I could not trust his lack of pain to rule out significant injuries. When CT scans revealed fractures in his pelvis, both shoulder blades, multiple ribs, and several spinal vertebrae, I assumed that his lack of pain was related to the meth of which his pupils spoke. After his urine test was positive for meth, he admitted that he had indeed shot up just before the argument, which also explained his nonsensical decision to jump out the window.

see patients on meth nearly every shift working in rural Pennsylvania, while colleagues in urban ERs see similarly gaping pupils resulting instead from cocaine and its derivative, crack, both of which are more available in urban centers. The type of drugs a ravaged community reaches for to soothe its turmoil impacts the sort of patients that are evaluated and treated in an ER—and what their eyes look like. Drug epidemics are geographic, and communities ravaged by different socioeconomic forces choose their dope differently—though opioids seem to have joined alcohol as the universal intoxicant of choice for people in both rural and urban regions of the United States, and from all walks of life.

Just as eyes are a window to deeper truths, working in an ER is a window into the socioeconomic stresses of a community and the ills of region. The only way to begin to help is to clearly see the truth and refuse to look away.

13

MUCUS

Cristian is eighteen years old, and the first thing he does every morning is deal with mucus. His ritual begins with inhaling three different nebulizer treatments in a row, each sending an aerosolized medication deep into his lungs to soften the phlegm accumulating in their dark, mucous hollows. He coughs incessantly during the treatments, sometimes enough to sprain the muscles in his chest and back. Next comes the vest, a wearable device that uses oscillating bursts of air pressure to pummel his torso with vibrations. He sits through twenty minutes of this brain-rattling routine in order to shake loose his phlegm before going to school.

The singular goal of this ritual is to get the mucus out. Cristian estimates that each morning he produces about two shot glasses' worth in varying shades of green, and he piously performs it every morning without fail—otherwise, he'll be winded while walking up stairs and cough incessantly throughout the day, disrupting his high school peers.

Cristian has cystic fibrosis (CF), a disease characterized by the world's stickiest mucus. Like all CF patients, he was born with a faulty genetic mutation, one that makes his lung mucus thicker and more tenacious than usual. In CF patients, mucus lodges in airways with detrimental results.

A slight change in the qualities of a person's mucus wreaks havoc on their body, and it dominates Cristian's daily schedule.

Mucus's most fundamental quality is its consistency, and this is what differentiates it from plain water. While water flows and drips easily and fluently, mucus oozes. While water droplets grab on to each other, easily melding edges to form larger drops, viscous mucus holds on to itself cohesively and resists all disturbances. With a tenacity that water lacks, mucus clings to surfaces, including the linings of our air passages, and it puts up a fight when we try to clear it out. The difference between water and mucus is the same as the difference between fruit juice and jelly, between simple salt water and bone broth, between gumbo before okra is added and after—mucus has body to it. In our own bodies, it appears in many forms beyond just lung phlegm, including snot, saliva, and vaginal discharge, with less noticeable amounts in stool and sometimes urine. But whatever we call its various manifestations, all are simply variations on the same theme of mucus.

As a medical student, I learned quickly that mucus holds a special place among bodily fluids. Healthcare workers, who deal with bodily fluids of every sort, seemed to have a particular distaste for mucus above all the others. I've heard countless doctors declare their disgust for sputum's chunky, gelatinous, sticky texture. Many nurses have told me they would prefer cleaning up a patient's bloody stool, even C. *diff*, to disposing of mucous secretions any day. When friends and family members wonder about my work as a physician, they often ask if blood grosses me out—I explain that blood is not the bodily fluid that typically grosses out healthcare workers. Mucus is.

Dr. John McGinniss is a pulmonologist at the University of Pennsylvania whom I met in medical school, and he describes being a pulmonologist the following way: "I do mucus all day, every day." Though he chose a career focused on this bodily fluid, he admits that it evokes a unique visceral reaction in many people because of its consistency, its distinctive thick bubbly sound, and sometimes its smell. "When you sit next to someone on the bus or plane," he said, "and you see them coughing and hear wet, rattling mucus, you think to yourself, *Oh geez, what disease am I going to get now?*" More than any other product of the human body, he said, people associate mucus in particular with illness. "And there's an emotion attached to it."

My own perception as a medical student quickly moved from one of disgust to one of wonder and appreciation. I realized that mucus is made by the human body for the same reason it coats the bodies of many animals, plants, and fungi throughout the natural world—for protection. A mucous covering safeguards snails and slugs, those creatures bathed in a slimy layer that leaves a silver sheen as they trudge across leaves and sidewalks. Their mucous coatings prevent them from drying out, and also fight off microbes like a shield. Rays, sharks, and tamarind seeds are bathed in a similar layer of lubricating and defensive mucus, as are several species of mushroom.

But unlike these other creatures, the human body is not coated from head to toe in mucus—instead it appears only in specific areas where an opening disrupts the body's otherwise continuous outer veneer of tough, dry skin. There are several of these disruptions, where skin folds in on itself to form a pocket, and all of them are clustered in the body's face, groin, and backside. Some of them, like the sinuses, have blind ends, but most, like the mouth, nose, vagina, and rectum, are passageways leading deeper into the body's anatomical cavities and tracts. Even the branching air passages that fill our lungs are just another of the body's pockets, though one that is more complex and manifold than most.

What all our bodily openings have in common is mucus. Unlike the desiccated crust of regular skin, the linings of these areas remain perpetually moist thanks to the steady production of mucus. Our various invaginations are like the human body's wetlands—while most of the surface is dry land, every once in a while, you come across a soggy patch. And unlike skin's varying colors and shades, everybody's soggy patches are lined by a universally deep-pink, blood-rich layer called mucous membrane, a layer named after its primary product and the thing we all have in common.

Our bodies require these perforations to serve as entrances and exits, as transition zones between the body's dry outside and its forever moist innards. But at the same time, they're constantly in danger: besides enticing people to put objects into them, which often leads to an ER visit when they cannot get them back out again, the primary threat is microbial invasion. While intact skin provides a layer of keratinized armor to fend off bacteria and other invading microbes, each disruption in skin's continuity is a vulnerable chink in the armor and a potential avenue for them to breach the body. Infectious microorganisms love nothing more than to bask in our dank openings, thriving and multiplying to their heart's content in the humid darkness offered by the human body's pink pockets.

Every microbe that attacks us has its predilections for certain spots. Yeast bloom in the vagina (and sometimes the mouth), while influenza and coronavirus prefer the nose, throat, or deeper into the lungs for their moistened revelry. Gonorrhea is the least picky of them all—it will take whatever mucus-lined pocket it can get into and regularly invades the urethra, rectum, and vagina. Sometimes gonorrhea climbs farther into both male and female genitals tracts, reaching all the way to the ovaries and testicles. I've even seen it infect my patients' eyes and throat.

Precisely because the body's many fenestrations are not easily guarded, mucus is essential. As a universal defense weapon and survival strategy, mucus flows outward from all of them in a steady, unending tide to keep microbes out—they'd have to swim upstream against a viscous current to get in. Keeping our membranes perpetually moist is also essential for maintaining their health and integrity, and mucus accomplishes this as lubrication with a staying power that plain water could never muster. Though mucus is often annoying and repulsive, it shouldn't be hated—instead, in the right balance, it is the key to how we stay healthy against an onslaught of invaders. And in healthy times, we make only enough to coat our surfaces with a thin veneer, the minimum needed to carry out its protective mission unnoticed.

Too much of a good thing can cause serious damage, a fact that CF patients like Cristian know well. In the lungs, mucus requires precisely the right consistency to be coughed up or cleared out by the same self-cleaning mechanisms that deal with aspirated food and inhaled dust. The mucus elevator functions by microscopic hairs lining the airways and constantly sweeping mucus along, dragging with it any detritus in need of disposal upward toward the throat. This crucial mechanism provides a hygienic flushing out of waste, with mucus as the mop. A similarly constant flow of mucus helps clean and maintain all of the body's other pockets, too, but consistency is key.

Mucus needs the right mix of water, salt, proteins, and carbohydrates in order to generate precisely the right amount of jiggle to do its job. The genetic mutation that causes CF causes a malfunction of microscopic salt pumps in the lining of the airways, which prevents salt and water from providing appropriate hydration to sputum. As a result, it becomes thick and sticky, and the normal housekeeping mechanisms fail. Mucus accumulates in the lungs, allowing virulent bacteria to take up permanent residence and leading to recurrent bouts of pneumonia that progressively worsen the ability to breathe. So persistent are these colonizing bacteria that Cristian's doctors instruct him to wear a mask whenever he comes to the doctor's office—something to which the rest of us have only recently become accustomed. A mask protects both him and other CF patients from trading lung microbes that, once colonized, they may never be able to be clear from their lungs again.

The risk of getting sick is a big motivator for Cristian to stick meticulously to his lung-clearance activities. He wears the vest four times a day and does his nebulizer chore a second time before bed, spitting the equivalent of another shot glass of green stuff into the small garbage can he keeps beside his nebulizer machine.

Thankfully, Cristian doesn't have any intestinal issues with CF—some patients suffer from an accumulation of the same thick mucus in their gut, which can completely stop up the bowel and require surgical removal of the clog. Pediatric surgeon Dr. Douglas Katz, my colleague in Camden, New Jersey, has performed this operation many times on CF patients, and he described to me the mucus that he extracts from blocked intestines as "a thick, tenacious epoxy—a vicious booger that sticks to everything."

As a pediatrics resident, I treated CF patients hospitalized for acute flares of their lung disease, which usually caused difficulty breathing, worsening lung function, and increased mucus production. The patients were usually teenagers, thin and sickly, and through my stethoscope, their lungs offered a cacophony of every pathological breath sound I'd ever learned. I treated CF patients with the strongest antibiotics in the hospital, the drugs of last resort that I rarely ordered for any other patients. There are good reasons to reserve the newest and broadest-spectrum antimicrobials for only the sickest patients with the most drug-resistant strains of bacteria, and CF patients like Cristian are it. His most recent admission last July came after a few days of coughing so much he couldn't sleep at night. But a few days in the hospital receiving the most powerful IV antibiotics brought his breathing and mucus back to baseline.

As an internal medicine resident treating adults, I rarely saw CF patients, because they usually do not live that far into adulthood. In past decades, the life expectancy in CF was in the twenties, but with recent improvements in care, specifically with the better techniques Cristian uses to clear mucus from his lungs, life expectancy has doubled to the forties. Cristian knows that mucus will eventually ruin his lungs no matter how meticulously he cleans them each day, and one day a double lung transplant will be his only hope of survival. If he ever received a new set of lungs, they would make mucus with a proper consistency and Cristian's life would be forever changed.

McGinnis and his pulmonology colleagues at UPenn have started using brand-new targeted therapies for CF that promise to further improve the disease's still grim prognosis. These state-of-the-art medications act directly on the defective salt pumps in the lungs of CF patients, and they abruptly correct the production of mucus. But they make patients nervous, McGinnis said. For those with CF, coughing up mucus has been an integral part of their daily lives since childhood, so when its accustomed flow becomes abruptly halted, people feel unsettled. According to McGinnis, they worry that, since mucus is no longer coming up, it must be lodged deep in their lungs, something that has always meant danger. For patients with particular genetic mutations—unfortunately, not the one causing Cristian's disease—these medications are fundamentally changing lives like no other therapy in the history of CF care.

Mucus from all pockets are the body's background effluents with which people live their daily lives. Some make more, such as people who are allergic to nearly everything and cigarette smokers whose lung mucus production has been permanently set into overdrive by the chronic irritation of tobacco smoke. Others make only minimal amounts, and every female has her own pattern of color and amount of vaginal discharge. Because of our profound intimacy with the products of our own bodies, we easily notice when something changes.

Even subtle differences are obvious, and it is often these changes that drive people to seek medical attention in the first place. A returning traveler who had visited the developing world might find mucus in their stool; parents seek help when they feel mucus rattling like the vibrations of a purring cat inside the chest of their coughing child and worry about pneumonia; and women often seek a diagnosis because of subtle changes to their typical pattern of vaginal discharge. The practice of medicine is often dominated by the task of evaluating mucus oozing from one or another of the body's orifices.

Like every other bodily fluid, mucus is useful to me as a diagnostic tool. I quiz my patients on the details of their discharges, asking about how it has changed, when they first noticed it, what time of day it is the most bothersome, if anyone else around them has noticed the same change, and more. Interpreting mucus to identify specific diseases is like interpreting all of the body's other fluids—clues of consistency, color, and quantity give me the answers I need to make a diagnosis.

When I perform a physical examination on my patients, a significant portion of the assessment consists of peering into their moist pockets in search of mucus. I shine a light to illuminate their depths, spelunking for a diagnosis. In sore throats, I push the tongue out of the way and instruct a patient to say *ah*—this vocalization forces the tonsils into view, where I can see if they are coated with a thick pus, a sign of strep throat (or, less commonly, gonorrhea pharyngitis). When peering into a patient's vagina, I use a lighted speculum to look for the cottage cheese mucus of yeast infections or the thick ooze of chlamydia, while STDs in men usually don't make so impressive a volume of discharge. Examining each orifice has its own technique, and I learned to be proficient in all of them.

Sometimes I need to extract mucus in order to run tests on it and figure out the diagnosis, and this often requires more finesse. In patients with respiratory failure who require ventilators to breathe for them, the breathing tubes snaking down their throats provide a passageway for air, but also a window into the world of lung mucus. While rotating through the ICU as a resident, I learned the satisfaction of suctioning. Under a nurse's guidance, I fed a catheter through the breathing tube of my patient until its tip disappeared down the trachea, and the suction's slurping sound showed I was close to a good sputum chunk. I slowly pulled back on the catheter, easing the tenacious blob up the patient's trachea. It put up a good fight, but I finally pulled out a wad of greenish-yellow phlegm flecked with brown spots. I felt victorious, as if I had just reeled in a giant fish that was finally vanquished and landed on the shore. Delivering a good gob of mucus from within a patient's body is among medicine's most satisfying procedures.

Another big part of practicing emergency medicine involves talking people down from their worst mucus-induced anxieties and fears. A change in mucus means that something is going on, but it's more commonly the result of a viral infection that will resolve on its own rather than a more concerning bacterial infection requiring antibiotics. The rattling mucus inside a child's chest is usually just a cold with post nasal drip, and green-colored mucus, contrary to what many of my patients think, usually does not mean anything more serious is going on.

Despite the inordinate amount of time I spend discussing mucus, and the fact that many of the treatments I prescribe are targeted to reduce it, to dry it up, and to stop its flow, mucus is not the villain. When microorganisms slip past our defenses and get a foothold, mucus becomes our primary weapon. Its flow quickens to flush away microbes, and its color and consistency change because mucus bears arms—enzymes, antibodies, and rampaging immune cells join in combat, defending us against microbes while inadvertently thickening the brew like chicken's feet added to a soup. When mucus becomes abundant, gelatinous, and colorful, it is a sign that the body is waging a battle.

Though we notice mucus the most in illness, it is always there coating our crevices, and it helps us heal from injuries too. When I learned which of my patients' lacerations require stitches and which heal just fine left alone, I found a stark difference between injuries to regular dry skin and injuries to mucous membranes. Skin lacerations often require fastening with stitches, staples, or glue to hold their edges together and facilitate regrowth. But lacerations to the pink mucous membranes heal almost magically in a few days, usually requiring nothing from me at all. Even quite large injuries to the inside of a mouth, inner cheek, tongue, or vaginal mucosa heal quickly, and I usually repair with sutures only the largest and most gaping defects. The body's mucus acts as a self-made healing salve that soothes these injuries and speeds the body's recovery. While most people do not think to lick their wounds as dogs commonly do, I often wonder if a dose of mucus would actually help them heal.

Everybody learns in grade school that the human body is composed mostly of water—roughly 60 percent in adults goes the statistic. I learned this fact as a child and understood that humans share their aqueous makeup with all other organisms. I imagined living things as plastic bags filled with sloshing, transparent, liquid water, even though my own body seemed much more solid. Years later in medical school, when I studied the body's construction under both dissecting scalpel and microscope lens, there was almost no plain water to be found. Tears and urine were the only fluids that seemed truly watery, with no hint of viscosity to them, but they flowed out from the body and were not actually part of its internal makeup.

Instead, the human body's insides are filled with an extensive range of more gelatinous textures. Even blood is slightly thicker than water—the old metaphor about the primacy of familial relationships is also true physiologically. With a protein content similar to egg whites, blood is thick enough to be whipped, and it is noticeably more syrupy than water in the way it drips and flows out of traumatized patients. And consistencies of the body only get more viscous from there: the sludge inside our eyeballs is akin to raw egg whites in consistency, while some internal organs feel like firm, but still jiggly, calf's-foot jelly. The gooey truth is that the human body contains textures far more varied than the simplified dichotomy of "flesh and blood"—there are as many different consistencies within us as in a French *fromagerie*.

Another thing we learn in grade school is that cells are the body's basic building blocks, and that we are constructed from them like buildings are made of bricks stacked upon one another. But in medical school, I finally learned about what lies in between our cells. Called the extracellular matrix, or ECM, the fibrous and gelatinous substance fills all of the body's interstices, surrounding cells like the body's packing peanuts enveloping fragile objects for shipping. If cells are the body's bricks, then this mucus-filled network running throughout the entire body is the mortar.

In contrast to the mucus that flows outward from the human body—the familiar stuff that we deal with day to day—the ECM is the body's internal mucus, which we never usually see. Yet it makes up much of us. In histology class, I looked at hundreds of microscope slides from every part of the body, and nearly every single view contained snapshots of ECM in one form or another. We typically focused on the cells in each slide, but in between and surrounding them was ECM. It appeared as loose strands of fiber—the rebar giving flesh its shape and tensile strength—with mucus providing the filler, like poured concrete providing structural support and resisting forces of compression.

Histologists call the ECM's mucus the "ground substance," as if to highlight its foundational and elemental role in the body. The ground substance and the ECM it comprises form a substantial portion of every body part and organ, accounting for roughly 15 percent of an adult's weight. As the body's largest single component, it integrates all our parts into a whole, and it is this continuum of mucus that makes us squishy.

Scientists are just learning the importance of the body's internal mucus, especially when it comes to building artificial organs that will one day (hopefully) be transplanted into patients in need, including lab-grown lungs for CF patients. Though stem cells get much of the media fanfare, it turns out that the ECM in which stem cells are embedded is equally important for them to function and grow properly. The extracellular mucus surrounding cells serves as the body's marketplace, where currencies of nutrition, waste, and messages of cellular communication are exchanged.

Like the body's outside mucus, its internal mucus also plays an important role in defense. When microbes invade the body—not just into the shadowy depths of a moist pocket but truly into our flesh—they must traverse directly through ECM, which can slow them down like squishy mud grabbing at a charging horse's hooves. The same goes for malignant tumors attempting to metastasize, which can move through mucus only by secreting specialized enzymes that break it down. When the body sends reinforcement white blood cells to help, the fight takes place in the same mucous milieu, meaning mucus is not only a sign of battle, but the battleground itself. The ground substance is like the dark matter of the human body—foundational yet forgotten, until recently. And a better understanding of it as the glue holding the body together may be one of the keys to discovering the medical treatments of the future.

It turns out that living beings are less like bags of sloshing water, as I imagined in grade school, and more like a stew. While water makes up most of it, water alone is thin and empty of the organic molecules from which organisms are built—the stew thickeners. Water alone is lifeless. Instead, it is the mucus in our bodies that—so long as it has the correct qualities and consistency—is the key to a healthy life. The "fact" that our bodies are made mostly of water is misleading. What we are actually made of is mucus.

14

FINGERS AND TOES

have never been more aware of my fingers and toes than I was during my first trip to Russia. I arrived in Saint Petersburg around New Year's Day when the temperature in the city was well below freezing. Every time I left the bubble of a building's warmth, my fingers and toes quickly became cold and numb, and the constant discomfort was ruining my trip.

I had gone to Russia soon after finishing college to intern at a social research center where I would be studying the impact of environmental organizations on the Russian forestry industry—medical school was not something I was yet considering. Within a few days of my arrival, I was invited to celebrate Russian Orthodox Christmas with two colleagues, Tonya and Vanya, a married couple, and their friends and family. They would be traveling to a cabin in rural northwestern Russia, in what sounded like the middle of nowhere. I eagerly accepted the invitation.

Together with Tonya, Vanya, and their two young daughters, Sasha and Masha, I took an overnight train followed by a long bus ride. We disembarked from the bus in the evening where the street dead-ended into a snowy forest. It was dark out, and the bus's headlights showed that our path through the forest had not been cleared of snow at all. My fingers and toes were already feeling chilly from the rickety, uninsulated bus, and the thought of heading into deep snow alarmed me. Vanya, who spoke no English and had a massive beard hanging down to his chest, plunged into the snowpack first, and Tonya pointed for me to follow next. The snow reached up to my mid-thighs, and Vanya and I took turns in the lead, shoving a knee into the untrammeled virgin snow ahead with each step, as Tonya, Sasha, and Masha followed behind.

After a short time, I realized—to my surprise—that my entire body was warm, even my fingers and toes. In fact, my gloves and boots were filled with sweat. It was the first time since I had arrived in Russia that my digits felt comfortable while outside, and it happened in—of all places—a dark forest at night with snow nearly up to my waist. When we paused to rest, I breathed heavily with exhaustion, plumes of vapor spewing from my mouth, and the cold quickly started creeping back in. Within a minute, my fingers and toes had cooled off and begun to bother me, but once we started trudging again, they promptly warmed up. As long as I kept moving, they stayed comfortable, and I was able to enjoy the winter beauty around me, with snow-shrouded evergreens pointing upward toward a crisp star-filled sky.

We came to a clearing in the trees, and I saw a lonely log cabin in the distance, a light in its window and smoke spewing from the chimney. Tonya and Vanya's friends had already arrived by skiing over the snowpack and had begun heating up the house. Between the fire they had built in the stove and the hugs and greetings I received from complete strangers, I felt relieved to be once again inside a warm bubble, and with all my fingers and toes intact.

The cabin had four small rooms arranged in a ring around a wood-burning oven that took up its entire center. The *pechka* was a brick-and-mortar cube that stood as tall as me, and just as deep and wide, with a small opening on its kitchen side for feeding in wood, and it served as the house's single source of heat. When I woke the next morning, from my bed I could see Tonya building the day's fire, her head wrapped in bright red cloth. I watched as she fed lengths of pale pinewood into the oven's sooty maw with a long wooden paddle. She shoved a piece of burning newspaper underneath the lattice she had built, and it quickly caught fire and glowed the same color as her head wrap, a life-giving warmth unveiled from within the wood's drab, lifeless colors.

The fire Tonya made would burn slowly throughout the day, its heat radiating outward from the house's center into the surrounding rooms, seeping into its every corner. The pechka even sent its warmth down a short dark hallway toward the relatively distant front door, the house's proverbial fingers and toes. These traditional stoves are well designed to extract every possible bit of heat from burning wood, and it kept the house cozy and warm despite temperatures outside plummeting down to negative 35 degrees Fahrenheit. The pechka's central placement paid homage to the preciousness of heat in a frozen world—and the precariousness of human life and limb (especially the digits at limb's very end) in extreme conditions.

After breakfast, I set out on cross-country skis to explore the rolling farm landscape around the cabin. It was colder than it had been in Saint Petersburg, and despite the comfort I'd experienced while trudging through the forest the night before, soon after leaving the house, my fingers and toes began to ache.

I tried every trick I knew to warm them up. I remembered an old adage about putting a hat on my head when my feet were cold, but even wearing the largest, most cartoonishly towering Russian fur hat borrowed from Vanya didn't help contain the heat. The hat's furry flaps protected my tender ears from the wind, but it did nothing to warm my toes. I tried vigorously stamping my feet and kicking my numb toes repeatedly against rocks and trees along the trail to drive warmth back into them. At first, when my toes were only slightly numb, a few good wallops helped temporarily, but once the chill crept farther into my boots, thrashing my toes became useless. The fight to keep my digits warm and comfortable was a losing battle against the overpowering Russian cold, though pummeling my toes provided a therapeutic way of venting my frustration.

I hastily retreated back toward the cabin, my toes somehow simultaneously numb and throbbing in pain, and I noticed the birch trees scattered along the walking path. Each one reminded me of a human body: a white vertical trunk stood like a spinal column, with thin boughs forking off it like outstretched limbs. And at the boughs' very ends dangled the tree's tiniest branchlets, completely caked in ice and snow. Through translucent ice, I could see buds on the trees' branch tips, and I knew that, despite being frozen solid at the moment, in spring those buds would still swell and unfurl into leaves and flowers as if winter had never happened.

I was jealous of the trees. Despite their frozen fingers, the taunting birches stood placid and silent, while I was being driven nearly into a frenzy by the cold's assault on the tips of my own body, where my arms and legs branch treelike into digits. If my fingers or toes became caked in ice like theirs, I knew they would never survive. In comparison with the resilience of the trees' woody flesh, my own body seemed terribly fragile.

I wondered if my misery was simply due to the cheap winter clothing I had brought from the States. Before I left, a relative had given me as a gift a pair of knitted gloves that looked warm but didn't feel that way once I got to Russia. Wind easily passed through their loose knit and chilled my hands. My boots also quickly proved their inadequacy in the Russian cold. Or perhaps the underlying problem was my body—I wondered if I had poor circulation and my body simply could not tolerate such frigid weather.

Once I got back inside the cabin, I held my hands directly against the heated pechka itself. My fingers began to defrost and the rewarming pain was excruciating, far worse than the cold itself. I was frustrated remembering how toasty my digits had been on the hike in—I knew it was possible to keep them comfortable, but still I resolved to stay close to the house every time I ventured out so that I could rush back to my thermal refuge if needed, the only way I could protect my suffering fingers and toes from the vast, unforgiving frigid world outside the cabin's bubble of warmth.

The following day, Vanya showed me a useful trick—when my fingers became cold, he explained through Tonya's translation, I should pull my hands out of their gloves and bury them deep inside my jacket. I tried it the next time I went outside and found that holding my fingers directly against the skin over my chest quickly thawed them, as if I had returned to the cabin and held my numb digits against the pechka itself. I often wished I could accomplish the same feat with my toes, but my lack of flexibility made it impossible.

Tonya also introduced me to the value of mittens. Instead of separating each finger, as gloves do, into their own individual bubbles of warmth, mittens group them together so they can share and help each other stay comfortable. This is equivalent to rewarming a hypothermic person by sharing a sleeping bag with another body, a last resort when there is no other heat source available. The heat shared by fingers in a mitten—or a pair of humans in a sleeping bag—is more than the thermal sum of their individual warmths.

When I learned the story of the human body years later in medical school, my experience in Russia made more sense. Fingers and toes have the most trouble staying warm in cold weather owing simply to the human form. The body buckles and involutes over and around our digits, creating our most convoluted corporeal topography. With a much higher surface area than flat expanses of the chest, back, or belly, fingers and toes easily lose heat to the environment. As a result, when cold assaults the human body, they suffer the most, as I knew well.

Our digits are also the polar opposite of internal organs—they are as far away from the body's central trunk as our flesh gets. Organs inside the chest and abdomen, especially the heart and liver, are the human body's primary heat generators, and they are responsible for keeping body temperature in a normal range. I learned in biochemistry class that the continuous churning of our most active organs releases heat as a by-product because of the imperfect thermodynamics of metabolism's chemical reactions. This excess warms up blood as it flows through our core, before carrying the heat outward into our extremities. But as the body's furthest hinterlands, digits jutting out from our trunks into the cold void have the most difficulty sharing the warmth.

During intense physical exertion, like traipsing behind Vanya through a dark, snowy forest, muscles must work hard to propel the body forward. In so doing, the muscles connecting the pelvis and upper thighs join organs as partners in heat production. My churning muscles warmed me up as a by-product in the same way that a car's engine, which combusts gasoline primarily for locomotion, becomes too hot to touch. My fingers and toes stayed comfortable in the forest because of the same side effect of metabolism's blessed inefficiency.

Digits have muscles inside them, but they are puny. No matter how repetitively they are exercised—toes curled and unfurled inside boots, or fingers squeezed into a fist and opened again—only minimal heat can be generated. Most medium-sized muscles that do the actual work of moving our fingers and toes are located farther up our limbs, in the calf and forearm, but they are also limited when it comes to keeping us warm. Only the biggest muscles of the pelvic girdle are large enough, and only the kind of vigorous and grueling activity as it took to get to the cabin can generate sufficient heat to keep fingers and toes warm.

A house designed to withstand the coldest temperatures requires a core radiating warmth outward, just like the human body. The pechka's location in the house's center was completely different from the wood-burning fireplaces I had known growing up in suburbia. Fireplaces in the United States are often in a corner and pushed up against the house's periphery, the equivalent of attempting to generate heat in the fingers and toes and having it radiate into the core. American fireplaces are mostly for romance and aesthetics, with much of their heat lost up the chimney, while in Russia, pechka fires are for survival. When Vanya taught me to place my fingers directly against my chest or stomach, it was a way of denying my body's inherently branched shape—I did not have to wait for blood to slowly trickle all the way through my limbs to heat up my digits.

Once I learned about blood in medical school, I realized that, of all the nutrients it carries to the body's remote geographies, warmth is among its most important deliveries.

When our bodies lose in their tug-of-war with the cold outside world, the result is frostbite, the actual freezing of human flesh, and our fingers and toes are the body parts most at risk. Though I left Russia with all my digits intact, many of my future patients have not been so lucky.

A man I treated in Alaska was riding across the tundra on a snowmobile in winter, when he slid down a stream's embankment, and the snowmobile's nose crashed through the ice. Both his feet were stuck under the water, smooshed between the ice and the snowmobile, and he couldn't extract them no matter how he struggled. Once his submerged toes began to rapidly cool beneath the stream water, his body preserved core warmth by constricting blood flow to his feet. This standard physiologic response to cold extremities prevented blood from getting cooled in his toes and then bringing the cold back into his core, which would have chilled his internal organs. This averted the senseless loss of precious warmth, and may have saved him from hypothermia, but it also doomed his toes.

By the time a search and rescue team pulled him out hours later, all ten of his toes were frozen solid. Over the following days, they turned black—the universal color of dead flesh—and they shriveled and fell off his body like a tree sloughs dead and dried-out branches. In such extreme situations, our digits become expendable and the body will shed them to save its more vital organs and, therefore, its life. And while a tree can regrow branches, frozen and dead human appendages can never grow back.

It's not only fingers and toes that are at risk of frostbite. All of the body's distant dangling bits projecting from the balmy core into the cold surrounding air are similarly in danger, including ears, noses, and penises. Penises, most unfortunately, suffer from the same hazardous topography as fingers and toes—their convoluted digitlike surface areas mean they are not easily kept warm, and most cases of penile frostbite occur when a steady stream of air cools the organ further, such as while jogging in chilly weather. Testicles, on the other hand, are safe, despite being the only internal organs to hang pendulously off the body. They rely on a temperature several degrees below the body's core temperature for optimal sperm production, and the scrotum hangs lower when it's warm and shrinks when it's cold, adjusting its closeness to the body's bubble of warmth to maintain a happy medium. Unlike fingers and toes, testicles are clever enough to retract and stay safe in frigid weather.

It was only after I finished residency that I realized the extent to which fingers and toes really are in danger. During residency, my daily bread was diseases of the internal organs—heart, lungs, kidneys, and liver—while the digits were mere peripheral characters with walk-on parts in a select few ailments. Finger and toe problems rarely require admitting patients to a hospital's medical wards, so I had minimal experience with their afflictions. I also learned little about trauma in my training, and finished residency ignorant of how to manage patients with bodily injury—both the life-threatening kind affecting the trunk and head, as well as the less severe forms that impact the extremities.

When I worked in urgent care after residency, I found that injuries to the digits came in an almost infinite variety. I encountered patients with partial amputations, deeply embedded splinters, dog bites, and crushed nails. Everything about the world outside our bodies seemed hazardous to our digits, the body parts most victimized by the innumerable mini-traumas of daily life.

Burned fingers and toes were typical. I saw one child who had knocked hot oil onto his hand, and came into the clinic with his fingers wrapped in strips of his own shredded T-shirt. I saw patients whose digits were bent into unnatural angles by dislocated and broken bones, and I learned to swiftly snap them back into place. A teenage boy came in after he crushed his big toe in an ATV accident—the tip of it was sticking straight upward from his foot, and its abnormal angle confused me until I numbed the digit with an injection of lidocaine and began trying to piece it back together. I realized the tip of his toe had been flayed open, partially detached, and twisted after he rode his ATV a little too close to a tree. He probably should have gone to an ER instead of my urgent care clinic, but I was selfishly glad he didn't.

Each affliction was something new and challenging, and I learned from them all. Many of them introduced me to mishaps of the digits that I had never even imagined before. While waiting for the results of an X-ray of a patient's injured finger or toe, I often studied up on how to correctly manage a condition I was seeing for the first time.

A whole new and fascinating world had opened up to me—the practice of everyday medicine, with a focus on the most everyday of body parts. When diagnosing the internal ailments that I was used to, like kidney failure, liver disease, and pneumonia, I typically needed to order blood and urine tests, as well as imaging studies, to indirectly evaluate organs hidden from my view. But I could directly examine fingers and toes with my own senses. It was much more hands-on, and as a physician, I felt liberated using my own fingers for something other than typing notes and orders into a computer.

As a side benefit to this on-the-job education, I could offer advice to friends and family members with their own everyday bumps and bruises. I knew the best techniques for splinting disfigured digits, the warning signs when someone needed to go to the ER right away, and the times when an X-ray could wait. My experience with fingers and toes in urgent care felt like the most practical medical know-how I had learned yet, and I felt like a more useful doctor than ever before.

make a living through my fingers, as does almost everyone. They perform so many essential day-to-day functions, which is part of their problem. We reach out and act on the world with our hands, our only body parts dexterous enough to grab and manipulate objects; fingers are imperiled as we slam doors, reach for hot cooking pots, punch through windows, and pet supposedly friendly dogs. Surgeons in particular depend on their fingers to perform intricate operations, and I have heard of some buying insurance policies on individual digits in case they get injured. Many people spend much of their days sitting in front of a computer, but even typing can be risky—overuse injuries like carpal tunnel syndrome cause pain and tingling in the fingers. Precisely because they are so useful, fingers are forever in danger.

Toes have their uses too—the big one especially helps with balance, and it pushes off as we walk. And because toes travel at the leading edge of every step we take, marshaling the rest of the foot, they easily get stubbed. Working in urgent care showed me that it doesn't take great force to break the tiny bones strung together inside them. Keeping our fingers and toes safe requires care and attention every day, just as it does during the coldest winters. And to make matters worse, fingers and toes are among the most sensitive parts of the body—they have some of the highest concentrations of nerve endings anywhere on the body's surface. So, all these everyday injuries really hurt.

I was particularly fascinated by injuries to the nails, and quickly learned what to do when they were sliced through, smashed to pieces, or torn off in part or in full. Nails are a unique aspect of our digits: as protective plates designed to deflect sharp blades and slamming car doors, they take the brunt of injury off the softer flesh over which they grow. Without them, our finger and toe tips would be whittled down by daily life, the water torture of mini-traumas. We lack the built-in weaponry that animals have in claws and talons, and while human nails can break skin, they mostly play defense. And they work, sometimes.

But nails, more than any other body part, open the human body up to another kind of harm—self-injury. In urgent care, I saw a variety of nail-biting mishaps, like abscessed fingertips and too much flesh torn off. I am a nail biter myself, and I know intimately the mysterious tic of the brain that makes people use their own teeth to brutalize their digits. Because I touch patients so often with my hands, I want my hands to look presentable, so I use bitter-tasting nail polish to stop myself from mangling them. Though nails are theoretically designed to protect our fingers and toes from injury, they simultaneously invite further abuse. Along with the beating they take from the rest of the world on a daily basis, our fingers and toes often bear the brunt of our neuroses.

The most important lesson I learned about managing the many afflictions that befall our fingers and toes was in assessing their temperature. One morning in urgent care, I evaluated a woman whose engagement ring had gotten stuck. She was pregnant, in her third trimester, and her entire body had progressively swelled for months. But that morning, her left ring finger had swelled more than the other nine, and had become a polka-dotted purple sausage. I touched her ring finger with my own hands and found it noticeably cooler than the others, giving me crucial information: blood flow to the digit was being interrupted. Her ring was doubling as a diamond-encrusted tourniquet, and it was threatening her appendage's survival. People like to use their fingers and toes for decorating the body, or announcing a betrothal, but sometimes our symbolic adornments pose a direct threat.

Checking a digit's temperature is a necessary first step in evaluating almost any problem of the fingers or toes, as well as injuries anywhere on the arm or leg from shoulder or hip down. As long as the digits at the extremity's farthest end are warm, it answers my most basic question about the health of human flesh—it tells me that blood is appropriately flowing to it. But this woman's finger was cold, and I needed to fix that right away.

I tried every trick in the book to get her ring off. I squeezed her finger inside my fist to push excess fluid out, and when I let go, it had shrunk and showed the imprint of my own hand. But her ring would not budge. I greased her finger with surgical lube—nothing. I tried a trick that I learned from YouTube: I wrapped her finger tightly with a string, and then slipped its end under the ring, but when I pulled the string to unravel it, the ring still would not move. It was time to use the ring cutter. I slipped the device under her ring to protect her finger, and began cranking the cutting wheel. In less than a minute, the ring's silver band had split, and once I pulled it apart and removed it, the swollen sausage began to deflate and turned from purple to pink. When I touched it again a few minutes later, it felt warm.

The link between our digits and the body's warm core works the other way around too—the temperature of a person's extremities is a diagnostic clue about critical problems with their internal organs. The coldest fingers I ever felt were in a woman who was bleeding profusely with postpartum hemorrhage. When I walked into her room, her face looked pale and her lips were barely pink—a sign that the amount of blood loss was critical. I introduced myself and shook her limp hand, and her fingers felt like ice—it is surprising how cold still-living human flesh can be when drained of its warming blood. She looked around the hospital room agitated and confused, her brain's delirium and her frigid fingers both demonstrations of the same cardiovascular collapse.

Her extremities told me that the pechka in her center was failing in the battle to keep her fingers warm, even against a mild room temperature, and she was rushed off to lifesaving surgery a few minutes later. We connect with other humans by touching our fingers to theirs, whether in greeting or in displays of affection, and the mingling of our most distant body parts somehow feels like connecting to the heart. The same simple gesture tells me something vital about my patient's core.

A human body, like a house, can never be heated completely uniformly. Body temperatures range widely from the warmth of our cores to fingers and toes, which are usually a few degrees cooler. In balmy weather, we don't notice the body's temperature differences, or the essential relationship between the body's trunk and its extremities, but in the cold, it becomes painfully obvious. Fingers and toes are bad at fending for themselves, and they are wholly dependent on the warmth circulated from the crowded core's bustling and metabolically active organs, the body's pechka. Digits are the most utilitarian of all body parts, and sometimes the most expendable, and they remind us of their usefulness when they are threatened and beaten by the outside world.

The one time I suffered mild, superficial frostbite was not to my fingers or toes but to my cheeks while snowshoeing across a frozen lake in Ontario. Like digits, prominent cheekbones also jut away from the body just enough to be at risk. As I snowshoed into the wind, I was not even aware of the waxy white color growing on my face. Suddenly, I felt someone's fingers on both my cheeks. Another person on the trip had noticed, and with his hands warmed from inside furry mittens, he reached out and thawed my skin.

15

BLOOD

As a resident, I heard rumors that live leeches were kept in the hospital pharmacy, but I thought it was a joke. I had never seen a leech before and knew of these bloodsucking parasites only as historical footnotes from the days when doctors knew almost nothing about disease. For centuries, they mistakenly thought all illnesses stemmed from imbalances in the body's four "humors," a collection of vague and unquantifiable ghostlike essences that supposedly flowed all through the human body. Out of the four—blood, yellow bile, black bile, and phlegm—an excess of blood was most often incriminated as the cause of everything from headaches to gout to psychiatric disease. The obvious solution to this problem was a therapy called bloodletting, wherein a lancet's sharp blade was used to slice into a patient's veins, and blood was allowed to drip out until the physician deemed enough had been removed for the humoral balance to be restored.

Leeches gathered from the wild and applied to a patient's body were a milder technique for achieving the same end. A denizen of lakes, ponds, and waterways throughout the world, leeches are worms resembling slugs that get all their nutrition by sucking the blood of whoever or whatever tromps through their habitat. Applying leeches to a patient's body and letting them remove blood was less painful than the lancet, and physicians of old could prescribe the proportionate number of leeches called for by a patient's specific diagnosis, something akin to counting milligrams of a medication dose. In the Middle Ages, leech therapy was so commonly prescribed in Europe that physicians were colloquially known as leeches, and medical textbooks were called leechbooks. Physicians and leeches formed a unique interspecies bond, a partnership for the healing arts founded on our totally wrongheaded obsession with draining the human body of its blood.

The humoral theory of disease and its practice of bleeding patients fell out of favor as physicians learned more about health and disease. Of all bodily fluids, blood is the most essential—by flowing to every part of the body, it provides the necessities that all our flesh requires to stay alive. Purposely draining it would cause anemia, in which the body has too few red blood cells to carry oxygen to tissues. Historians of medicine postulate that up until the twentieth century, a patient was more likely to be harmed than helped by a physician and this was probably due in large part to our bloodletting.

To me, leeches symbolized the barbarism and ignorance of the bad old days of medicine. So when I kept hearing the same rumor about leeches being available in the hospital's pharmacy, I needed to see for myself if it was true. I called the pharmacy—as I had often done for advice on the dosages or timing of medications—but this time I asked the pharmacist if the rumor was true.

She chuckled. "Yes, we do have them down here."

The thought of a living animal among the pills, powders, and vials of injectable medications I was used to prescribing for my patients intrigued me. The hospital pharmacy held some of the most advanced and powerful cures known to humankind—chemotherapies, immunomodulators, and antibiotics—and it seemed incongruous for a living animal to be kept among them, a parasite harnessed by physicians specifically for its singular appetite for human blood.

"What specialty uses them?" I asked. The answer: plastic surgery.

I immediately contacted the on-call plastic surgeon, Dr. Edward Kobraei, to learn more. Though most people know plastic surgeons for their work reshaping noses and breasts, their basic job, as Kobraei put it, is to "move tissue around." They might relocate muscle and skin from one part of the body to another to reconstruct deformities left by chronic wounds, burns, or surgery. But for any transferred tissue to survive in its new location—for it to "take"—the proper flow of blood is essential, and this is where leeches can help.

I asked if I could observe a session of leech therapy, but unfortunately, Kobraei had no patients receiving it when we spoke. Two days later, however, he called me back—a patient had come in the night before with a severe finger injury, and Kobraei had initiated a regimen of leeches to save the finger. He told me the patient's name, Michael, and his room number. I rushed up to the plastic surgery ward.

Blood is the internal fluid most responsible for keeping the body's lights on—it is precious, and a terrible thing to waste. Every piece of our bodies, every bit of flesh, requires a constant flow of blood as the bare minimum to stay alive and continue functioning. Accomplishing this task is the entire purpose for the cardiovascular system—it serves as the irrigation system for the body, with blood as its business end. And when the delivery of blood is interrupted, even briefly, our cells begin to wither and die like plants rooted in soil parched by the lack of life-sustaining water. The heart is so central to bodily health, and the cessation of the heartbeat in cardiac arrest is the most time-sensitive emergency in all of medicine, precisely because blood ceases to flow. The death of a human body is actually a trillion microscopic cell deaths, and each is killed most proximately by the bloodstream's own arrest.

Blood is also our most complex bodily fluid—it serves as a universal transport medium, distributing every conceivable nourishment throughout the body. Blood moving through the intestines picks up nutrients absorbed from food and conveys them to every far-flung cranny of flesh. It similarly flows through the lungs, where it picks up oxygen for distribution, and this is where blood turns red—it is rich in iron, and when exposed to oxygen, it turns the same color as a piece of metal rusting in the air outside. New-borns turn pink when they finally start breathing for the same reason: rust-colored blood begins flowing to all of the body's tissues, where its oxygen is released to hungry cells. Blood then un-rusts back to blue before returning once again to the lungs to re-rust in a lifelong, continuous cycle.

Because blood has a smorgasbord within it, including protein, carbo-hydrates, fats, salts, minerals, and cells, it bears a comprehensive overview of a human body's health. This is why physicians regularly do blood work as part of their evaluations. Analyzing blood can reveal almost everything about the body, from the state of its organs to the balance of hormones to the presence of a hidden infection somewhere inside. Results from blood work are a secret code, its numeric ciphers telling physicians whether a per-son is healthy or ill, as well as precisely how dire the situation is. We are reliant on blood work for diagnosing and monitoring many diseases, a more advanced relationship between doctors and the blood of their patients than in the Middle Ages. This new form of bloodletting, phlebotomy, is mainly for diagnostic purposes, rather than therapeutic ones, and it offers unparal-leled insight into a patient's hidden insides.

For the same reason that blood is useful to physicians—its wide-ranging composition—blood is also extremely nutritious. While in the womb for the first nine months of gestation, a growing fetus's only source of nutrition is mother's blood, providing everything it needs to grow and develop. Even for adults, animal blood provides well-rounded nutrition, and it is a staple food in many cultures—especially among natives of the Far North, where the vi-tamins and minerals it contains are hard to find anywhere else in the natural world. While riding a bus in Kamchatka, Russia, I chatted with an Orthodox priest who had recently spent time proselytizing among the Koryak people of the North—he complained that the indigenous families refused to follow his teaching not to drink the blood from their reindeer, an age-old custom.

As he spoke, I thought of the savage hordes of mosquitoes I had experi-enced during the Kamchatka summer—they drain people of blood drop by drop, stealing precious iron from the body. I imagined that drinking animal blood was a good supplement to replace the parasitized nutrients. Blood is one of the most complete foods in the world, which is why it can serve as the sole food for mosquitoes, as well as for ticks, bedbugs, some vampire bats, and, of course, leeches. More than any other bloodsucker, however, leeches offer the most convenient medical therapy: they draw a decent amount of blood with each meal, they cannot fly away, and they are easily stored for months in a refrigerator. This is what originally endeared them to the medical profession, and why they are still used today.

Michael had begun his day like any other, with no hint that it would quickly turn into a bloodbath. A construction worker from Vermont, he was in his kitchen drinking coffee and looking out the back window of his house. He noticed that the family dog had gotten one of its hind legs tangled in its leash, so he walked out toward the large maple tree where the dog was tied up. He unclipped the leash from his dog's collar and reached down to free the leg. Just then, the dog took off at full speed—a loop of leash encircled Michael's thumb like a noose and instantly tightened down to nothing.

His thumb dangled with a suddenly skimpy affiliation to the rest of his body, and blood poured from his hand. At that moment, a medical odyssey began that would blend the most sophisticated surgical techniques in the world with some of the most ancient.

Michael's wife rushed him to the nearest ER, where doctors placed tightly wrapped pressure bandages on his hand to stop the bleeding and injected painkillers to reduce his searing pain. As the ER doctor inspected the damage, it was immediately obvious that a specialty surgeon would be needed to reattach Michael's thumb, and he was transferred to MGH in Boston.

Kobraei learned about the injury while Michael was still an hour away, en route via ambulance. He was waiting in MGH's ER, dressed in standard blue scrubs and a matching cap as the paramedics wheeled Michael in on a stretcher. Kobraei gingerly unwrapped the bulky gauze dressing covering the all-but-amputated digit. The thumb was cold to the touch, its color the ghostly gray that flesh turns when drained of its blood. He looked closely at the wound and could see shredded blood vessels, and he knew that Michael's thumb would not survive if blood flow was not quickly reestablished.

Michael was brought to the operating room, where Kobraei and a team of surgeons donned sterile gowns and headlamps with magnifying lenses. With the first few surgical incisions, the thumb did not bleed at all, a sign that its shredded arteries were unable to deliver any blood. Kobraei repaired both the arteries and the veins by sewing them with thread almost too thin to see with the naked eye, and only then did bright red blood begin to well up in the gashes and drip onto the blue operating room towels below. Bleeding is something only healthy living flesh can do, a perverse proclamation of vitality.

Over an hour later, the operation was finished, and Michael was wheeled to the plastic surgery ward. When Kobraei visited a few hours later to inspect the initial results, he found the thumb in mixed condition—it was warm and full of blood again, demonstrating that the shredded arteries had been effectively repaired. But the thumb was badly swollen and had a purplish hue, which meant that the veins had failed.

According to Kobraei, arteries are easier than veins to stitch back together, their walls being thick and sturdy, woven with muscle fibers that allow them to withstand the rippling shock wave of every heartbeat. Veins, on the other hand, are thinner, floppier, and more fragile—they often cannot stand up to the rigors of tightly pulled sutures. Even when veins are reconnected in the operating room, blood's slower flow through them makes it more prone to clotting, just as slow-moving or stagnant water freezes sooner than rushing streams.

Blood was flowing into Michael's thumb, but it was unable to flow back out again—the other side of the circulatory equation for all flesh. The finger was getting congested with too much blood, and Kobraei knew the best medical therapy for the situation: leeches.

met Michael in his hospital room the following morning as he awakened to the sound of his nurse's voice. The tired, glazed look in his eyes spoke of fragmented sleep, narcotics, and exhaustion from the pandemonium that had enveloped his life the moment that a loop of leash performed its savage surgery. I introduced myself and asked for his permission to watch, which he readily granted.

The nurse carried into the room a red plastic container secured with a latch. Without her saying anything, Michael extended his arm and placed his injured hand onto the bedside table—he was clearly accustomed to the routine. His swollen thumb jutted upward out of a jumble of bandages covered in dried maroon blood. Two thin metal pins poked out from the thumb's tip—they helped stabilize the shattered bones and repaired blood vessels, holding his thumb in an ironic thumbs-up. The nurse placed the red container on the bedside table next to Michael's acquiescent hand.

The "leech motel," as the nurses call it, housed the day's installment of twenty-four leeches; one was applied to the thumb every hour. It had arrived on the plastic surgery ward early that morning, hand-carried from the hospital pharmacy. More typical medications are usually transported from the pharmacy to hospital wards through a pneumatic tube system, but living creatures like leeches are too delicate for that method's violent, throttling suck. The leeches needed to arrive on the ward healthy and hungry— after all, a leech's appetite is its therapeutic potency.

Lowering her head toward Michael's hand, the nurse took a close look at the thumb. She pressed her gloved finger against a pink area near the thumb's tip, and I saw the flesh momentarily blanch a grayish white before quickly filling up again with pink. This refilling of the smallest capillary blood vessels was a reassuring sign that blood was reaching his threatened finger, and also an indication that the leech would eat well.

The nurse unlatched the leech motel and picked up a pair of slender tweezers. After fishing around in the plastic tub, she pulled the tweezers back out, and grasped between their shiny metallic arms was a squirming ribbon about three inches long. The leech was much more colorful than a slug—a rich tapestry of yellow and olive-green patches speckled its back, and a beige stripe ran down each of its flanks. The leech flopped and craned its body in a futile attempt to escape the nurse's tweezers.

With a tight grip, she held the leech directly up against the tip of Michael's thumb. The leech's squirming became more animated—I imagined that it had caught the scent of blood. Michael watched with detached apathy as the leech's mouth sucker explored his skin. The nurse softly whispered words of encouragement to the writhing creature.

Finally, the leech found its spot and planted its mouth onto the thumb tip. It suddenly became motionless as the first taste of blood satisfied its frenzy. I asked Michael if it hurt—he said he hardly felt anything. As the leech's body began to ripple with peristaltic waves of guzzling, Michael's head dropped back onto the pillow.

The nurse and I returned to Michael's room fifteen minutes later, when, she estimated, the leech would be about finished with its meal. We walked in to find Michael asleep and the leech, still feeding on his thumb, more than twice as large as when we left, swollen taut with blood. The nurse pulled out her tweezers and grabbed the leech just as it let go of Michael. Her timing was perfect—the previous day, she recounted, she had returned to the room to find a blood trail snaking from the foot of Michael's hospital bed across the white linoleum floor into the bathroom. She followed the blood trail and found the leech motionless behind the toilet. I wondered if it had sought out a dark, watery retreat similar to its native habitat as a place to rest and digest its meal.

I leaned over to look closely at Michael's thumb—its swelling had improved significantly with the leech's feeding, and its color had gone from purplish to bright pink. Since his surgery, Michael's thumb had suffered from precisely the problem that medieval physicians mistakenly saw in their patients' entire bodies—too much blood. Congested tissue does not heal properly, and leeches offered another way of removing excessive blood until the damaged veins could regrow over the next few days.

The hourly blood ritual continued steadily for three days, and then Kobraei's team slowly weaned the schedule over another two days as the thumb's veins came back online. Leech therapy made Michael anemic (as it did legions of patients throughout history), but his primary medical issue was one of the very few in which blood needed to be removed, even at the expense of the rest of his body. A few blood transfusions easily cured his anemia. Leech therapy was eventually stopped completely, and the thumb was able to handle its blood flow on its own. Michael was discharged home to Vermont for months of healing and physical therapy.

When I spoke to him six months later, Michael was back at work, and back to skiing the mountains around his home. He could hold on to a ski pole with no trouble, but his thumb didn't tolerate the cold so well as it once had. Blood coming to warm his thumb would need to navigate the scarred course through his vasculature forever after, but at least he still had a thumb.

Leech therapy works because it removes blood from congested tissue, but the key to its effectiveness is not actually in the blood swallowed by leeches. When a leech finishes its meal and detaches, the wound keeps on bleeding, a fact about leech bites that physicians and naturalists have recognized since ancient times. As traditional leech therapy became obsolete in the early twentieth century, this observation kept leeches medically relevant and indelibly changed how physicians practice.

The knowledge revolution that created modern medicine was sparked when pathologists conducting autopsies discovered that the true seats of illness are found in tumors, infections, and organ failure, rather than in humors. They also discovered the paramount importance of blood clots in killing people. Blood must clot, of course; otherwise, every paper cut and nosebleed would be life threatening, but clots forming inappropriately within the bloodstream were found to be the cause of many of the most common and most fatal diseases, including heart attack, stroke, and pulmonary embolism. This new understanding spurred scientists to search for medications that could treat and prevent blood clots, and they found one in leeches.

Intensive research into leech bites led to the discovery that leech saliva contains a powerful blood thinner called hirudin. For leeches, hirudin prevents their blood meal from clotting, which lets them feed for longer. After they detach, the leftover saliva in the bite wound keeps blood flowing, which accounts for a more significant volume than the five milliliters that a medical leech swallows during an average meal. The extra drainage accounts for the bulk of leech therapy's effect when used by plastic surgeons like Kobraei. For humans, an understanding of leech saliva led to the first commercially available blood thinner in the early 1900s—an extract made from ground-up leech heads.

Today, much of my job as an internist consists of thinning my patients' blood to prevent and treat clots, and there is an ever-increasing variety of blood thinners from which to choose. Derivatives of hirudin are still widely used, and just within the ten years since I finished medical school, there are several new blood thinners on the market. Physicians have more tools than ever before with which to treat the myriad disorders caused by blood clots, but leeches bestowed on our profession the very first.

This should not be surprising—parasites make the perfect source for medical treatments. By definition, they make a living at the expense of their hosts, and are able evade the human body's defenses and feast off its nutrients, turning our weaknesses into their own nutrition. Having lived and evolved alongside humans and other animals for eons, drinking the blood of generation after generation, leech saliva developed its unique ability to interrupt blood's normal clotting function. Every time a leech sank its teeth into a human throughout history, its saliva swirled together with human blood, a mixing of human and leech essences that epitomizes the dance of parasite and parasitized. This was the laboratory in which leeches developed their pharmaceutical ability to manipulate our physiology, exploit its vulnerabilities, and extract from us what they need.

And physicians make a living in precisely the same way. Every medication I prescribe to patients is similarly designed to alter a patient's biochemistry, to manipulate physiology in hopes of restoring health or alleviating the symptoms of disease. When I prescribe blood thinners to my patients, I am tinkering with their blood just as leeches do. The tools of my trade are not that different from the tools that parasites use—sometimes they are exactly the same thing. And in the years to come, we may find that live leeches can help cure an even wider variety of illnesses.

When I think about medicine's future, I often wonder what revolution will change the way healthcare operates today. Which aspects of today's practice will be considered barbaric? There are sure to be many. My best guess is that future physicians will look back and marvel at how physicians of the early twenty-first century continued to drain human bodies of blood, even a century after the humoral theory's death. Physicians like me are guilty of drawing too much blood from our patients—we often order unnecessary tests, and those hospitalized for prolonged periods can become anemic from the blood draws alone. When, in the future, an entire battery of tests can be run on a single drop of blood, we will be seen as brutish, and still all too leech-like.

The practice of medicine always strives for improvement, especially by learning from the natural world, though its history is littered with wrong turns. Sometimes the latest life- or limb-saving advancement springs from past mistakes, and sometimes long-abandoned treatments from the past crop up again. Physicians will forever look to the natural world for guidance on how to heal the human body, just as we first learned how to manipulate human blood from a creature that had already spent eons perfecting its craft.

Acknowledgments

Many people over decades helped make this book a reality. I am grateful to those who taught me most about the human body, beginning with the man who became my cadaver and continuing through every patient I have ever evaluated.

Several doctors significantly impacted my medical career, including Dr. Larry Weisberg and Dr. William Surkis, who first introduced me to the big picture of being a doctor; Dr. Jim Withers, who inspired my career choices; and Dr. Jack Preger, who showed me the immense good that one doctor can do.

I've had many mentors and guides in my explorations of the natural world, but none so important as Marc Gussen—he taught me to identify wild edibles and make bows, but he also showed me a deeper understanding of how humans fit into the natural world. I am thankful to Maria Tysiach-niouk, who taught me how to travel, thereby changing the course of my life, and to Larry Millman, who fed my fascination with geographic extremes, odd foods, and travel writing. Others who have broadened my horizons include Leslie Van Gelder, Dr. Stephen Moorman Tom Brown Jr., Antonina and Ivan Kuliasov, Herman Ahsoak, Steve Brill, Jim Riggs, Matt Richards, Kielyn and Dave Marrone, Craig George, Glenn Sheehan, Anne Jensen,

Dr. N. Stuart Harris, Dr. Warren Zapol, "the Leech Lady," Dr. Tom Hennessy, Dr. Kahabi Isangula and Dr. Vivek Krishnan, Dr. Sunny Jain, and all the medical students I met in Mumbai.

In the writing of this book, Sandra Bark provided invaluable assistance with editing and structure. I am grateful to Lauren Bittrich for embracing this project and seeing it through, as well as to Noah Eaker, Sarah Murphy, Megan Lynch, and Bob Miller at Flatiron Books for believing in this book. Thanks also to my literary agent, Jeff Silberman, who first pushed me to embark on this book. I was fortunate to receive useful feedback on the manuscript from Eli Kintisch, Brett Mole, Vivian Reisman, and Drs. Benjamin Yudkoff, Tamar Reisman, and Daniel Flis.

Writing this book involved cold-contacting a wide variety of people and peppering them with weird questions about oddball topics. I offer heartfelt thanks to everyone who gave me their time: Ari Miller, Kiki Aranita, Iris Katzner, Sandra Mole, Eyglo Geira, Cynthia Graber, Nicola Twilley, Dr. Neal Yudkoff, Yuval Ben Ami and Jennifer Kirby for discussing unusual foods; Drs. Buddha Basnyat and Ken Zafren for teaching me high-altitude medicine; Drs. Evangelia Bellas, Libby Hohman, Edward Kobraei, David Dinges, John McGinnis, Lee Kaplan, Douglas Katz, Tomasz Stryjewski, Kahabi Isangula, Sriram Machineni, Paul Janmey, Hossein Sadeghi, Mousa Younesi, Charlotte Nussbaum, Daniela Kroshinsky, and Sarah Gilpin for indulging my atypical inquiries about feces, fat, mucus, and more; and Richard Blakeslee, David Weight, Ani Choten, Theodore Ruger, Owen Paterson, and Drs. Ankur Kalra, Nimesh Vesuwala, Yulia Slonova, David Barnes, Tommy Heyne, Bernard Kinane, and Marjorie Bravard for humoring my unusual questions about everything from plumbing to sleep in hospitals to religion to food law. Thanks also go to Marilyn Zeidel for insight into aspiration, and to Thaneswar Bandhari, Sonia Beker, and Dr. Aliva De for helping facilitate interviews.

For insights into Jewish law, I am thankful to Yehuda Wexler, Dr. Shaya Wexler, Chaya Wexler, and Rabbi Shlomo Wexler. In searching federal records pertaining to lungs, I received assistance from University of Pennsylvania law librarian Susan Gualtier, as well as Kirstin Nelson and Ashly

Johnson of the USDA. Susan Cavanaugh at Cooper Medical School also helped with historical references, and Lauren Steinfeld and Helen Oscislawski provided legal advice.

And finally, to Anna, the primary reason that this book ever even came close to seeing the light of day. The words on this page could never adequately express my gratitude to her for being my support, my editor, my life partner, and my muse. She kept me grounded, focused, and reasonable, as she has with everything else I've done in life since we met. I am eternally grateful for her unending love and support.

References

THROAT

13 *But even when delivered . . . causing choking or pneumonia:* Finucane TE, Bynum JPW. Use of tube feeding to prevent aspiration pneumonia. *Lancet.* 1996 Nov 23;348(9039):1421–1424. James A, Kapur K, Hawthorne AB. Long-term outcome of percutaneous endoscopic gastrostomy feeding in patients with dysphagic stroke. *Age Ageing.* 1998 Nov;27(6):671.

17 *Research shows that . . . forgo aggressive therapies:* Periyakoil VS, Neri E, Fong A, Kraemer H (2014). Do unto others: doctors' personal end-of-life resuscitation preferences and their attitudes toward advance directives. *PLoS ONE.* 2014;9(5):e98246.

FECES

45 *In India, I was shocked . . . Indian kids every year:* Lakshminarayanan S, Jayalakshmy R. Diarrheal disease among children in India: current scenario and future perspectives. *J Nat Sci Biol Med.* 2015 Jan–Jun;6(1):24–28. Million Death Study Collaborators. Causes of neonatal and child mortality in India: nationally representative mortality survey. *Lancet.* 2010 Nov 27;376(9755):1853–1860.

GENITALS

56   *Studies of the moon's phases . . . female genital tracts:* Pochobradsky J. Independence of human menstruation on lunar phases and days of the week. *Am J Obstet Gynecol.* 1974 Apr 15;118(8):1136. Gunn DL, Jenkin PM, Gunn AL. Menstrual periodicity: Statistical observations on a large sample of normal cases. *J Obstet Gynecol Br Emp.* 1937 Oct;44:839.

56   *The purported menstrual synchrony . . . inconsistent finding:* Ziomkiewicz A. Menstrual synchrony: Fact or artifact? *Hum Nat.* 2006 Dec;17(4):419–432. Yang Z, Schank JC. Women do not synchronize their menstrual cycles. *Hum Nat.* 2006 Dec;17(4):433–477.

PINEAL GLAND

78   *And research has shown . . . the body's circadian rhythm:* Gabel V, Maire M, Reichert CF, et al. Effects of artificial dawn and morning blue light on daytime cognitive performance, well-being, cortisol and melatonin levels. *Chronobiol Int.* 2013 Oct;30(8):988–997.

81   *Health outcomes of . . . cardiovascular disease, obesity, and diabetes:* Hoevenaar-Blom MP, Spijkerman AM, Kromhout D, Verschuren WM. Sufficient sleep duration contributes to lower cardiovascular disease risk in addition to four traditional lifestyle factors: The MORGEN study. *Eur J Prev Cardiol.* 2014;21(11):1367. Knutson KL, Van Cauter E, Rathouz PJ, et al. Association between sleep and blood pressure in midlife: The CARDIA sleep study. *Arch Intern Med.* 2009 Jun 8;169(11):1055. King CR, Knutson KL, Rathouz PJ, Sidney S, Liu K, Lauderdale DS. Short sleep duration and incident coronary artery calcification. *JAMA.* 2008 Dec 24;300(24):2859. Lao XQ, Liu X, Deng HB, et al. Sleep quality, sleep duration, and the risk of coronary heart disease: A prospective cohort study with 60,586 adults. *J Clin Sleep Med.* 2018 Jan 15;14(1):109. Sabanayagam C, Shankar A. Sleep duration and cardiovascular disease: Results from the National Health Interview Survey. *Sleep.* 2010;33(8):1037. Patel SR, Hu FB. Short sleep duration and weight gain: a systematic review. *Obesity* (Silver Spring). 2008 Mar;16(3):643–653. Cappuccio FP, Taggart FM, Kandala NB, et al. Meta-analysis of short sleep duration and obesity in children and adults. *Sleep.* 2008 May;31(5):619. Spiegel K, Tasali E, Penev P, Van Cauter E. Brief communication: Sleep curtailment in healthy young men is associated with decreased leptin levels, elevated ghrelin levels, and increased hunger and appetite. *Ann Intern Med.*

2004 Dec 7;141(11):846. Greer SM, Goldstein AN, Walker MP. The impact of sleep deprivation on food desire in the human brain. *Nat Commun.* 2013 Aug 13;4:2259. Cappuccio FP, D'Elia L, Strazzullo P, Miller MA. Quantity and quality of sleep and incidence of type 2 diabetes: A systematic review and meta-analysis. *Diabetes Care.* 2010 Feb;33(2):414–420.

86 *Much about its . . . impacts the immune system:* Guerrero JM, Reiter RJ. Melatonin-immune system relationships. *Curr Top Med Chem.* 2002 Feb;2(2):167–179. Besedovsky L, Lange T, Born J. Sleep and immune function. *Pflugers Arch.* 2012 Jan;463(1):121–137. Spiegel K, Sheridan JF, Van Cauter E. Effect of sleep deprivation on response to immunization. *JAMA.* 2002 Sep 25; 288(12):1471–1472. Cohen S, Doyle WJ, Alper CM, Janicki-Deverts D, Turner RB. Sleep habits and susceptibility to the common cold. *Arch Intern Med.* 2009 Jan 12;169(1):62. Rechtschaffen A, Bergmann BM, Everson CA, Kushida CA, Gilliland MA. Sleep deprivation in the rat: Integration and discussion of the findings. *Sleep.* 1989 Feb;12(1):68. Rechtschaffen A, Bergmann BM. Sleep deprivation in the rat: An update of the 1989 paper. *Sleep.* 2002 Feb 1;25(1):18–24.

86 *while other studies . . . tumor-fighting properties:* Bartsch H, Bartsch C. Effect of melatonin on experimental tumors under different photoperiods and times of administration. *J Neural Transm.* 1981;52:269–279. Lissoni P, Chilelli M, Villa S, et al. Five years survival in metastatic non–small cell lung cancer patients treated with chemotherapy alone or chemotherapy and melatonin: A randomized trial. *J Pineal Res.* 2003 Aug;35:12–15.

BRAIN

92 *As undesirable . . . young, fit trekkers:* Honigman B, Theis MK, Koziol-McLain J, et al. Acute mountain sickness in a general tourist population at moderate altitudes. *Ann Intern Med.* 1994 Apr 15;120(8):698. Hackett PH, Rennie D, Levine HD. The incidence, importance, and prophylaxis of acute mountain sickness. *Lancet.* 1976 Nov 27;2(7996):1149–1155.

94 *Altitude also impacts . . . people more irritable:* Shukitt-Hale B, Lieberman HR. The effect of altitude on cognitive performance and mood states. In: Marriott BM, Carlson SJ, eds. *Nutritional Needs in Cold and In High-Altitude Environments: Applications for Military Personnel in Field Operation.* Washington, DC: Institute of Medicine (US) Committee on Military Nutrition Research, National Academies Press; 1996. www.ncbi.nlm.nih.gov/books/NBK232882.

95   *At altitude, malfunction . . . memory, and decision-making*: Pun M, Guadagni V, Bettauer KM, et al. Effects on cognitive functioning of acute, subacute and repeated exposures to high altitude. *Front Physiol*. 2018 Aug 21;9:1131.

FAT

133   *The average American . . . pounds every year*: Dutton GR, Kim Y, Jacobs FR, et al. 25-year weight gain in a racially balanced sample of U.S. adults: The CARDIA study. *Obesity*. 2016 Sep;24(9):1962–1968.

139   *He pointed . . . than other populations*: Fumagalli M, Moltke I, Grarup N, et al. Greenlandic Inuit show genetic signatures of diet and climate adaptation. *Science*. 2015 Sep 18;349(6254):1343–1347.

LUNGS

145   *In 1969 . . . called human food*: 9 C.F.R. § 310, 325 (lungs).

146   *But for humans . . . lactating udders*: 9 C.F.R. § 310.17 (mammary glands).

151   *In the past . . . bang for the buck*: Shulchan Aruch, Code of Jewish Law, Yoreh De'ah, 39:1. Found at: www.sefaria.org/Shulchan_Arukh%2C_Yoreh_De'ah .39?lang=bi. Accessed August 18, 2020.

FINGERS AND TOES

186-7   *Penises . . . jogging in chilly weather*: Hershkowitz M. Penile frostbite, an unforeseen hazard of jogging. *N Engl J Med*. 1977 Jan 20;296(3):178.

BLOOD

193   *Out of the four . . . gout to psychiatric disease*: Leech. In: Britannica, T. Editors of Encyclopaedia. *Encyclopedia Britannica*. Britannica Website. http://www .britannica.com/animal/leech. Accessed August 18, 2020.

194   *In the Middle Ages . . . called leechbooks*: Magner LM. *A History of Medicine*. 2nd ed. New York: Informa Healthcare; 2007.

194   *Historians of medicine . . . helped by a physician*: Wootton D. *Bad Medicine: Doctors Doing Harm Since Hippocrates*. New York: Oxford University Press; 2007. Kang L. *Quackery: A Brief History of the Worst Ways to Cure Everything*. New York: Workman Publishing; 2017.

202   *After they detach ... an average meal:* Mode of Action. Leeches USA. http://www
       .leechesusa.com/information/mode-of-action. Accessed August 25, 2020.

202   *For humans ... leech heads:* Nowak G, Schrör K. Hirudin—the long and
       stony way from an anticoagulant peptide in the saliva of medicinal leech to
       a recombinant drug and beyond. A historical piece. *Thromb Haemost.* 2007
       Jul;98(1):116–119. Jacobj C. Verfahren zur Darstellung des die Blutgerin-
       nung aufhebenden Bestandtheiles des Blutegels. D.R.P., 1902; Patent Nr.
       136103, Klasse 30h/204. Jacobj C. Über hirudin. *Dtsch Med Wochenschr.*
       1904;33:1786–1787.